Digital Transformation in Contemporary Times

A Panorama for Reshaping and Reimagining the Future

Digital Transformation in Contemporary Times

A Panorama for Reshaping and Reimagining the Future

Dr. Ashima Saxena
Dr. Priyanka Chadha
Mr. Shivam Gaur
Dr. Ajay Bhatia

NAMYA PRESS.com

Published by

Namya Press

NAMYA PRESS.com

India: 213, Vardaan House, 7/28 Ansari Road, Daryaganj, Delhi-110002
USA: 10685-B Hazlehurst Dr. # 31013 Houston, Texas - 77043
Email: namyapress@gmail.com
Website: https://namyapress.com

Edition	:	First Published in 2022 (Namya Press)
Title	:	Digital Transformation in Contemporary Times: A Panorama for Reshaping and Reimagining the Future
Editors	:	Dr. Ashima Saxena, Dr. Priyanka Chadha, Mr. Shivam Gaur, Dr. Ajay Bhatia
ISBN	:	978-93-5545-098-2
Copyright	:	© Dr. Ashima Saxena, Dr. Priyanka Chadha, Mr. Shivam Gaur, Dr. Ajay Bhatia, 2022 All rights reserved

Preface

The currently ongoing 4th Industrial revolution has been unprecedented in terms of scale, scope, and complexity, being primarily characterised by explosiveness of its development and the disruptiveness of its technologies. Technology-led disruption of business models, impacting people and organizations across cultures, geographies, industries and economies, is slowly becoming the order of the day. There is hardly any economic or commercial activity across the globe that has or is not going to get impacted by the rampant pace of digitalization-led change

Digital transformation encapsulates the integration of digital technology into all areas of a business resulting in fundamental changes to how businesses operate and deliver value to their customers. Technology spending and digital transformation endeavours have been on the rise, but majority of such change programs are either not succeeding for want of the right digital mindset, or, lack of leadership and operational rigour, or, even, non-availability of budget in tough economic conditions perpetuated by the regional tensions, supply chain constraints, Covid pandemic and frequent natural disasters, etc. Many times, it is treated in a siloed manner or just like any other IT program, or does not involve the confluence of people, process, technology, culture and leadership, due to which it becomes destined for failure.

Digital transformation is simply a case of rapid adoption, transformation and performance, but if overlooked or ignored, organization gets firmly entrenched in the decline stage. While certain industries or organizations have been early adopters, there are many which are lagging and still moving at a very slow pace to catch up with the change.

Efforts have been made in this book to examine six different industries and business verticals to explore the extent of digital transformation, critically evaluate the positives along with the emanating challenges, suggest solutions, and point out future directions and areas of further research.

Chapter 1 explores diverse aspects of digital transformation across industries. It observes that while digital transformation is a boon to the human race, it could also turn out to be a curse if not handled with care and with responsibility. Chapter 2 organises and summarises the research and conclusions regarding the use of digital business models to improve government networks and services. This study examines how and whether the e-government initiatives of the central government in India have been reworked to be more user-friendly especially after the pandemic struck; and, if these models are flexible enough to incorporate ongoing economic changes. Chapter 3 endeavours to break the myths around "Digital mindset", clarify why it is relevant and virtually a "make or break" condition for any organization envisioning their future in this digital age. The chapter also details out well-researched key principles and common threads that constitute a digital mindset, and finally how should an organization go about systematically fostering a digital mindset. Chapter 4 highlights the importance of focusing on the organizational culture and employee wellbeing in the digital era. Technology by itself can do little unless the people who work with it feel part of something special and valued. For change and innovation to thrive, organizations require agility, leadership, decision making, while individuals need to be aware of their thoughts, feelings & behaviours, and approach changes with a sense of belongingness & clarity.

In the legal domain, Chapter 5 points out that with the advancements in technology, concerns of privacy violation have increased tremendously in the era of digital banking. The chapter highlights some legislations and draws reference to landmark judgements to trace the evolution and reformation of the legal stance on data protection and privacy. Chapter 6 explores the benefits and challenges of the present digital court systems in India. While transparency, accessibility, storage/preservation of court records, agility, witness/litigant security, etc are some of the benefits, however, the challenges of complicated processes, viz., lack of techno-legal expertise, lack of connectivity, fund crunch and other logistical issues, etc., are still dragging the system down. The chapter makes certain recommendations to overcome the challenges and leverage the full potential of digitalisation. Chapter 7 explores the applicability of copyright legal framework and patent laws on the "AI-created" works, and calls for re-examination of copyright standards for AI systems all over the world. Chapter 8 observes that that as the trend for e-Commerce has picked pace in recent times, so have the disputes and conflicts related to e-commerce. As such, there is a dire need to explore Alternative Dispute Resolution mechanisms, including online arbitration. But several challenges, including awareness and expertise necessary to execute online arbitration still exists, before it could be considered as a truly feasible and convenient method in place of the earlier ways of seeking dispute resolution.

In the realm of Ecommerce, Chapter 9 discusses the usefulness and drawbacks associated with the use of digital payments in commercial transactions in the Indian economy, and gives suggestions for making digital payments mechanisms more effective and secure. Chapter 10 delves in length on the Quick commerce as a business model, which is here to stay and is catapulting a lazy model of convenience shopping to the new normal of an urban lifestyle. It goes into the evolution, technology, operational process being followed, along with future directions, to meet the rising customer expectations for quick, reliable and efficient delivery of groceries through eCommerce platforms.

Chapter 11-14 explore the impact of digitalisation in Banking & Financial sector. Chapter 11 traces the evolution of Fintech industry and the gradual emergence of Digital Banking Units (DBU) in near future. It discusses the various areas of operations under the Fintech industry and the challenges being faced by the regulatory bodies in governing them, besides how these developments are impacting the traditional banks and financial institutions. Chapter 12 provides detailed insights about the three important aspects of digital finance in India i.e., enormous development of digital finance in India, the challenges and hindrances being faced, and the future scope of digital finance in India, including suggestions to overcome the hurdles and achieve its full potential. Chapter 13 gives an outline of digitalization and digital finance, highlighting the developments being triggered by the new FinTech revolution and how the incumbent suppliers of banking and financial services are undergoing a significant transformation due to quickly shifting economic climate, as well as onerous and expensive regulatory requirements of digital finance. Chapter 14 discusses the commercialization of the internet in India, the critical aspects to be considered by an organization when it embarks on its digital journey, the concerns and challenges in the global context, and future scope of online commercialization in India.

Exploring the higher education industry, Chapter 15 reviews the literature thoroughly to provide a scholarly background on the adoption of e-learning technology and resources in higher educational institutions, including the contributing factors, challenges faced, and provides key suggestions for effective implementation of e-learning in higher education institutions. Chapter 16 delves on the aspect of internationalization of education, digital education and issues relating to regional integration, relevant to the overall sustainability of modern education as envisaged by NEP 2020. Global practices and importance of research and development in the realms of digital and internationalization of education have also been emphasized. Chapter 17 discusses how Covid-19 pandemic has played a significant role in dramatic and quick shift in the Indian Education landscape, brought through digitalization and online education. The chapter further highlights certain key issues and challenges which need to be overcome while integrating online learning into the traditional educational system.

Chapter 18 discusses how artificial intelligence technology is bringing about significant changes in the solar energy sector and paving a new way for energy generation and storage, apart from better forecasting, management of grids and scheduling maintenance, etc. It also discusses certain harmful side effects, which need to be managed proactively.

Chapter 19 examines the diverse factors and trends of digitalization, and the dynamics due to inter-relations among these factors, which are acting as a catalyst in the development of the hospitality and tourism industry and opening up a new array of possibilities for the industry.

This book provides a comprehensive view on the diverse aspects of digital transformation that have been instrumental in redefining and reimagining the business models, organizational design, delivery and customer experience of different business/industry sectors. It points out areas of digitalization that are currently posing challenges, offers suggestions for improvement, along with pointing out areas for future research.

—Editors

Contents

BANKING & FINANCE SECTOR

HIGHER EDUCATION SECTOR

SOLAR ENERGY SECTOR

HOSPITALITY AND TOURISM SECTOR

SECTION - 1
Business Organization

1

Digital Transformation - Boon or A Curse

Sampriti Phukan
Assistant Professor, School of Law, Manav Rachna University

Abstract

Human Beings have survived through the test of time because we are considered as the superior beings compared to the other beings. This is evident from the time when we were dwelling on trees and in caves, since then we have been developing new and innovative ways to make our lives better and comfortable. Technology is the product or the process developed by the intellectual human brain. Without technology we could not have been alive. However, every coin has two sides of it and the discovery and invention of technology is no exception.

Digital transformation has made human lives very convenient in every sphere. Be it medical field, engineering field, education, banks or even our day to day basic activities like buying vegetables or booking a table in a food joint. It has given a new dimension to everything that surrounds us. However, there is a question whether this change is for good or is it leading to something so disastrous that we are unable to foresee it as we are so engrossed in our luxuries of life provided by this digital transformation. This digital transformation that we see is making us more compassionate or is it taking us to a zone where there is no existence of human emotions. This chapter is discussing whether this digital transformation that has gulped society is a boon or is it a curse to human civilization.

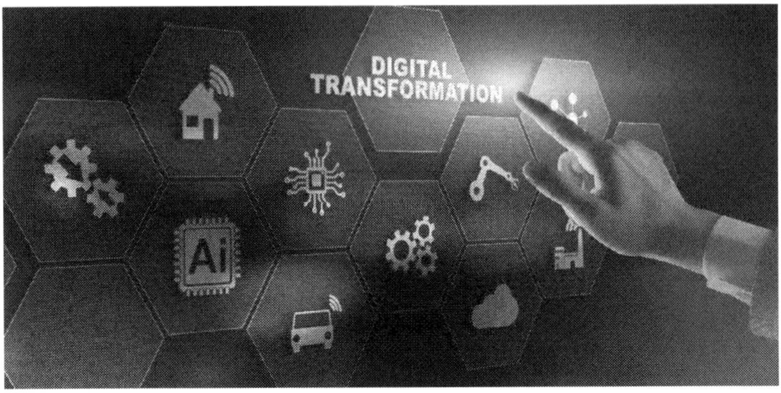

Photo Source: Google Images

Keywords: *Digital Transformation, development, human civilization, technology, boon and curse*

Introduction

When the society was formed it was very basic and simple. People struggled to make their lives comfortable, safe and secure. However, gradually with the changing time things started changing and humans, the superior of all the creatures started making discoveries. There were numerous discoveries and inventions taking science and technology to different heights. Inventions and discoveries led to many new dimensions in society. It also led to the Industrial Revolution which opened the doors to many more opportunities to the whole world. The pace with which people and the society was functioning suddenly accelerated. Everything suddenly started changing rapidly.

With this development everything changed for man; his relationship with nature, society and family. Everything changed so rapidly that no one even released the change. The Scientific Revolution changed the basic characteristics of human beings. People's blind faith in religion and nature saw the light of logic and rationality. Thus, with the advent of the scientific revolution interpretation of everything started its foundation on logic, rationale and reasoning. There is evidence that the Indus Valley Civilization (World's oldest civilization) also had scientific technologies used in house buildings, drainage systems, agriculture and various tools made for different activities of day to day chores. It didn't stop there, with the changing time more and more advanced technologies came into existence which changed the whole paradigm of the society and its existence. However, one of the most important discoveries that have taken all the individuals to a different horizon all together was the advent of the digital or the virtual world for all of us. We are in the 21st century where technology has grown into the most dependable and essential things for the whole mankind.

History of Digital World

The human race in thrust to make life and livelihood easily accessible and convenient has been always working towards making new discoveries. The discovery of technologies to produce machines, radios, televisions, computers etc., is known as electronics. The use of these electronic systems that uses the numbers 1 to 0 to record sound or store information and that gives high quality results is termed as DIGITAL. "Digital Transformation (DX) is about adopting disruptive technologies to increase productivity, value creation, and social welfare".[2] With the advent of the digital technology the whole dynamics of man with society and nature changed forever. It made life faster paced and advanced. The whole pattern of working changed. It made things more productive in a very short span of time. This brought a huge revolution in the history of mankind. This discovery enabled man to take a giant leap forward many more discoveries such as digital medical equipment, digital education system, digital art (cinema, art, music etc.), business of various items and everything under the sun. This created a huge impact on the human race. Earlier everything was done by man manually and it was very time consuming and tiring. The productivity of man was restricted. However, with the development of the digital world everything became fast paced. Man was able to do many tasks in a very short span of time and this led to a revolution in every sphere of life and business.

"Digital transformation requires strong leadership to drive change. But it also requires a vision for what parts of the company you want to transform. Companies in all industries and regions are experimenting with and benefiting from digital transformation. Whether it is in the way individuals work and collaborate, the way business processes are executed within and across organizational boundaries, or in the way a company understands and serves customers, digital technology provides a wealth of opportunity. Focus is important as no company is transforming all the areas at once. But the best-managed firms are constantly identifying new ways to redefine the way they work in the new digital era. The use of technology to radically improve performance or reach of enterprises

— is a hot topic for companies across the globe. Executives in all industries are using digital advances such as analytics, mobility, social media and smart embedded devices as well as improving their use of traditional technologies such as ERP to change customer relationships, internal processes and value propositions. Other executives, seeing how fast digital technology disrupted media industries in the past decade, know they need to pay attention to changes in their industries now."[3]

Present Day Scenario

The present scenario of this digital era is intertwined and complicated. There is hardly any field felt where the touch of this digitalisation has not reached. However, it has touched every individual on this planet through some way or the other. The blessings this digital transformation brought in could be seen and felt during the pandemic of COVID-19. The blessing of this Digital Transformation was very evident in the medical field, educational field, and essential commodities. It won't be wrong if we say that digital transformation was the only knight in shining armour of mankind during the COVID-19 pandemic. There was another pandemic which affected the world 100 years back- Spanish Influenza. However, back then the treatment, education for children and other basic functioning of life was different compared to the COVID-19 pandemic. It is all due to the invention of Digital Transformation in the present day.

Medical Science: The most vital requirement for the whole world as a community during this pandemic was medical science. "Over the past two decades, many hospitals and health systems have adopted digital technologies in their various functional areas. In many cases, however, they took a piecemeal approach to numerous initiatives—from installing electronic health record (EHR) systems to building apps to trying disruptive technologies such as artificial intelligence (AI)—while remaining largely focused on the same business and customer models. The COVID-19 pandemic significantly altered this status quo for the health care industry. Virtual health and care delivered in the home became the model of not only necessity but also preference. But this change was not as sudden as it might look. The pandemic was an accelerator of several trends, including shifting consumer preferences, rapidly evolving technologies, newer talent models, and clinical innovation. In the face of these trends, as hospitals and health systems work toward adapting their businesses, a well-defined approach toward digital technologies will likely be at the core of this transformation strategy". [4] Even though there are numerous benefits of the digitalization of the medical sector one cannot ignore the fact that there are many issues or concern that needs our attention. Medical history or issues related to health is always a matter of privacy for every individual. When this private information leaks out or becomes accessible to others, it violates the right to privacy under article-21 of the Constitution of India. Every individual has the right to privacy guaranteed under this article and infringement of the same leads to the violation of the fundamental rights.

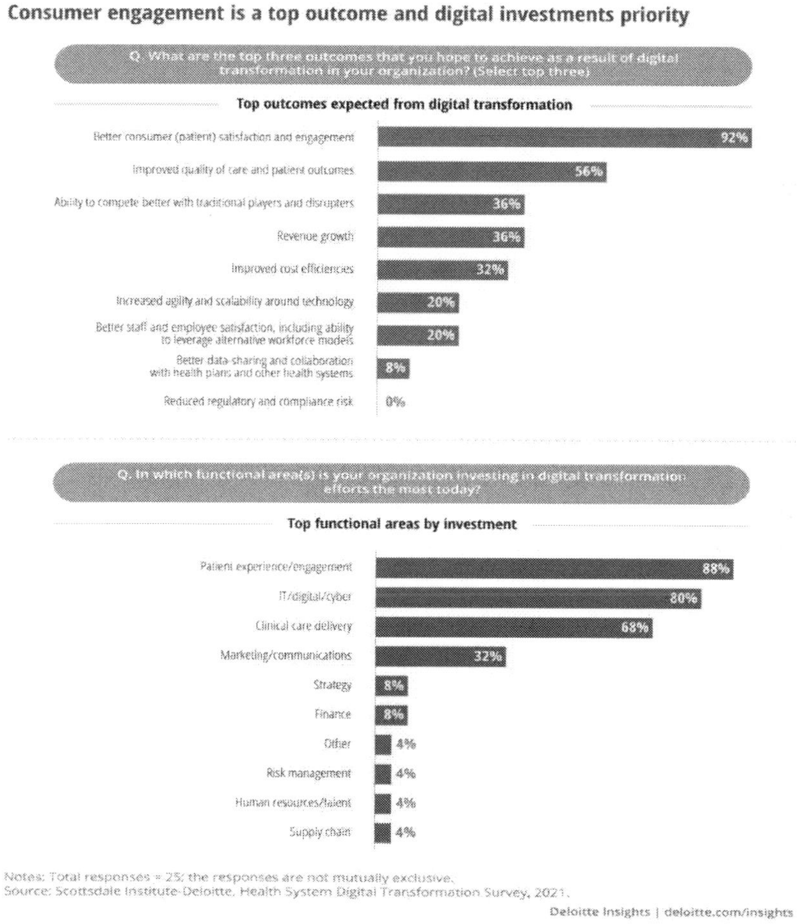

Source: (https://www2.deloitte.com/us/en/insights/industry/health-care/digital-transformation-in-healthcare/_jcr_content/root/responsivegrid_380572564/advanced_image.coreimg.95.800.jpeg/1635891147809/us164823-fig1.jpeg)

Thus this digitalization of every detail of the individuals is a genuine threat to their identity, personal space and liberty. "The National Digital Health Mission (NDHM), according to the government, is the first step in strengthening India's health system and moving it toward the United Nations Sustainable Development Goal 3: achieving universal health coverage. It aims to digitize the health records of all individuals — clinical tests, illnesses, medical prescriptions, and reports — under one health identification.

The proposed "womb to tomb" health identity, while aiming to improve efficiency in patient care and ease of access to health records, has also raised fears of it being used as a mass surveillance tool. The proposed "womb to tomb" health identity, while aiming to improve efficiency in patient care and ease of access to health records, has also raised fears of it being used as a mass surveillance tool. A group of public health researchers who studied the impact of linking health schemes with the national identity number, Aadhaar, concluded that while efficiencies of incentivizing use of health care services improved, there were a number of serious ethical challenges including interruptions in the treatment of patients with tuberculosis and acquired immunodeficiency syndrome as patients feared a breach of their confidentiality."[6]

Article 12 of the Universal Declaration of Human Rights and Article 17 of International Covenant on Civil and Political Rights states that, "No one shall be subjected to arbitrary interference with his privacy, family, home or correspondence, or to attacks upon his honor and reputation. Everyone has the right to the protection of the law against such interference or attacks". In an ever upgrading society, the privacy of people has become a prey to technological developments and advancements. Like a human attitude has two extremes, these developments can be a boon or bane to the society. When a person's information becomes available to anyone around the world, it can lead to piercing his privacy. In this way, his basic human right of leading his life with dignity can be infringed. In the current scenario, the privacy invasion trend has turned its head on medical data. Nowadays, the medical information of patients is stored by the hospitals digitally for betterment of its services. But such information is not impregnable at all times. Furthermore, in some countries, this information is accessible by the employers of such patients and their medical insurance companies as well. This makes their Personal Health Information (PHI) available to umpteen numbers of people. Any unscrupulous use of such PHI for other purposes outside this sector can be detrimental and scandalous to an individual. Hence, there is an urgency for reinvigorating the existing laws for protection of medical data privacy [7]."

Educational field: When technology has touched everything in this world from medical to business, education is no exception. Technology brought a new revolution to the whole concept of education during the pandemic all over the world. It made learning more concept based and interesting for students compared to traditional methods of teaching and learning of education all across the globe to impart knowledge to children. The development of the new age technology of online classes, lectures, tests, webinars, conferences and connection with each other was a total game changer. It gave the students an edge and gave them a new direction to go and explore. Through this technology they were able to connect with their teachers and classmates.

"Today, technology enables forms of communication and collaboration unlike in the past. Students in a classroom in the rural U.S., for example, can learn about the Arctic by following the expedition of a team of scientists in the region, read scientists' blog posting, view photos, e-mail questions to the scientists, and even talk live with the scientists via a videoconference. Students can share what they are learning with students in other classrooms in other states who are tracking the same expedition. Students can collaborate on group projects using technology-based tools such as wikis and Google docs. The walls of the classrooms are no longer a barrier as technology enables new ways of learning, communicating, and working collaboratively. Technology has also begun to change the roles of teachers and learners"[8]. In this new era of technology the words like laptops, tablets, smart phones, smart watches, 4G/5G networks are no alien words. They have become an integral and vital part in everyone's life. Technology and digitalization has taken over all the nooks and corners of human lives. "When the pandemic caught us unaware in March 2020, followed by the continued lockdowns, we all realized how digitisation saved the day by altering our day-to-day lives, into what we started calling the 'new normal!' Though the country had already been undergoing rapid digitisation long before the pandemic hit us, the lockdowns and social distancing guidelines just accelerated the pace of this transformation, as it was the only available alternative. The ramified impact of the Covid-19 pandemic all through its three waves reiterated the fact that evolutions and change are the only constant and will continue to remain so. The pandemic-induced socio-economic turmoil impacted everything we can think of, and education was no exception. As per data by the UNESCO Learning Portal 2020, over 87% of the world's student population was affected by lockdowns and quarantines, and 1.52 billion students were absent from school and other educational institutions. The Covid-19's abruptness, uncertainty and volatility compelled the educational system to adapt quickly in tandem with the new learning landscape. To be more resilient, equitable and inclusive in these trying times, educational institutions leveraged technology like never before, tried to address learning losses, make amends to minimise the impact on vulnerable and disadvantaged communities. UNESCO established the Global Education Coalition, to mobilise and support learning continuity. Today, it counts 175 members working around three central themes: Gender, connectivity, and teachers" [9].

"Digitization has no doubt changed our education system, but we cannot say that it has diminished the value of our old time classroom learning. Neither do we want something so priceless to turn into dust. The best part about the digitization of education in the 21st century is that it is combined with the aspects of both; classroom learning and online learning methods. Walking hand in hand both act as a support system to each other, which gives a stronghold to our modern students. Digitization in education has also proved to be the right method for saving resources. Online examination platforms have restricted the frivolous usage of paper, directly confining the cutting down of trees. This way the digitization of the education industry in the 21st century proves to be a boon to our society" [10].

Banks/business: When everything around us changed its dependency and trading to digital platform how can the banking sector be left behind. This shift from manual to digital banking process initially was considered a hiccup as not many were comfortable using it in accordance with the latest technology. However, this digital literacy among India is a welcoming change. "India's digital payments landscape has transformed dramatically over the past five years on account of rapid expansion in digital infrastructure, UPI-led migration to digital, pandemic-led acceleration of shift in customer preferences, growing merchant acceptance network and disruptive innovations by fintechs. The habit of carrying your wallet while making purchases may soon be replaced by carrying your phone, as consumers continue to go cashless, findings of a recent study suggest. According to a report jointly released by Boston Consulting Group (BCG) and PhonePe, 65 percent of all transactions, or two of every three transactions, in terms of value, are expected to be digital by 2026. India's digital payments market is at an "inflection point" and is expected to more than triple to $10 trillion by 2026 from the current level of $3 trillion, the report said. This growth would be driven by increase in merchant payments, it added, as more and more merchants shift to QR (quick response) based payments in offline sales. The findings of the report exclude B2B (business-to-business) and G2B (government-to-business) transactions ". There is no denying the fact that when the business or entrepreneurship has gone digital how can money transactions be in traditional mode. This has led to many crimes such as credit card frauds, account hacking, fake online shopping apps, email spoofing, phishing etc., making life more complex and insecured. If there is a good and positive side to digital transformation then these are the dark and the negative sides which cannot be ignored. The authorities are also struggling on their part to fight these menaces. However, the rate with which its increasing is not countered with equal force. There are a number of cases rising where the shopping websites are cheating people with poor quality products and money. These practices are causing anxiety and trust deficit among the public and the real businesses are getting hampered. In these kinds of crime the consumer has limitations in regard to filing of cases in the district consumer forum due to lack of evidence and the person/company against whom they want to file cases are mostly untraceable. This makes them handicapped in taking action against the perpetrator.

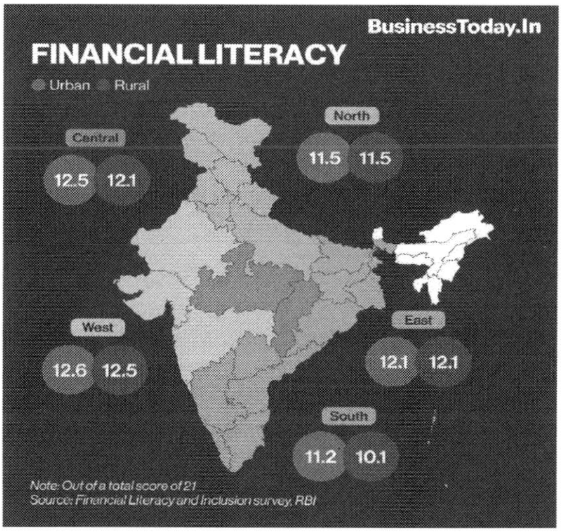

Source: https://www.businesstoday.in/latest/economy/story/digital-transactions-grow-by-88-in-three-years-315406-2021-12-13

Legal Sector: Apart from all the development in various fields the legal field has also been touched down by digitalisation. The conventional court rooms are transforming into digital courtrooms. The hardbound case files, journals, digests and case laws are changing into soft copies. This has made the access to various important documents accessible to the public and advocates all around the world. The options for e-filing of the case and applications, e-hearings, contact-less meetings and conferences, virtual courtrooms etc., have changed everything in the legal fraternity. It has made everything every organised, error free, easily accessible, time saving and convenient for everyone. There is a section of the society who is illiterate and is technologically challenged. When the court proceedings become digitalised it becomes nearly impossible for them to access its services. Anyways people are not comfortable with the court proceedings and making it digitalised adds up the uneasiness. There is another issue that is being faced by the advocates and the public is the unavailability of the internet connections in the remote areas of the country. Even though we talk about the new age of digitalisation of the whole country, it's a distant dream as most of the places are still not able to access the internet.

Conclusion and Suggestion

It has been rightly said by **John Wooden** that "*there is a price to be paid for achieving anything of significance. You must be willing to pay the price.*" There is nothing in this world that comes for free. It is so relevant when it comes to digital transformation. This brought a revolution in the history of mankind forever however at a cost. There is no limit for the desires and aspirations of human beings. There was an alarmingly high rate of crimes in the digital world. The alarmingly high rate in the crime rate done through the digital tool compelled the law making authorities to think and enable a legislation "The Information Technology Act, 2000". It's a well known fact that whatever discovery has been made by man has both positive and negative outcomes. Likewise digitalisation too has positive and negative impact on the society as a whole. There is no denial that it is the need of the hour and everything is dependent on it. We cannot think of development in any field without digitalisation. However, there are many disadvantages of digitalisation for younger and older generations of people. Digitalisation has made people grow into isolation and loneliness. People have become addicted to these gadgets and it has led to the social isolation and sedentary lifestyle compelling people to get away from each other. It has affected them mentally as well as physically. The discovery of the new age technology has made work more accurate and productive resulting in very rare chances for human errors. Human emotions and feelings are evading due to this rapid digital transformation. It has also made more and more people jobless as their work is taken over by the digital platform as the percentage of human value is declining at an alarming rate. The dependency and addiction to the digital platform is also a matter of concern as it is leading to many psychological and behavioural changes in people around the world. It is also leading to a more sedentary lifestyle of humans whose repercussions are evident from their rising health issues. Further it has also led to development of habits like procrastination. Due to the easy access of the contents on the internet and untraceable nature of the internet there is an alarming rise in the crime and terrorism. Infringement of privacy is one of the major concerns and worries in this digital transformation era.

One of the main reasons for the increasing crimes in the digital sphere is lack of awareness about it. Over the years people have seen and witnessed traditional crimes and they know what can lead to it and how to evade it. However, this is the new genre of crimes about which people are not much aware about. Thus, it is making it easier for the perpetrators and difficult for the general public to evade it. Awareness camps and workshops should be conducted in schools, colleges, universities, workplaces etc., about these rising crimes in the society and how to prevent it. The authorities should also be trained and capacity building should be done about how to handle these crimes and the criminals. The law should be made more stringent in order to deter the criminals. The younger generations should be taught and encouraged to take up real hobbies, sports, book reading etc., in order to divert their attention from the digital world to the real world. Younger and older generations should be made aware about these crimes and the dangers associated with them.

Thus, in retrospect we should realise that when digital transformation is a boon to the human race it can also be a curse if not handled with care and with responsibility. We all should never forget the quote of Christian Lous Lange that, "*Technology is a useful servant but a dangerous master.*" There are two sides of this digital transaction- good and bad, however it depends on which one we are feeding more.

References

Christof, E. and Carlos, Henrique C. Duarte (2018), Digital Transformation; Software Technology Digital Transformation, Research Gate https://www.researchgate.net/profile/Carlos-Henrique-Duarte-2/publication/326241618_Digital_Transformation/links/5b4d14f3aca27217ff9b05e4/Digital-Transformation.pdf <last visited on 28.7.2022>

Westerman, G., Bonnet, D., & McAfee, A. (2014). The nine elements of digital transformation. *MIT Sloan Management Review*, *55*(3), 1-6 (https://dixital.cec.es/wp-content/uploads/presentacions/presentacion08.pdf,)

Appleby,C., Hendricks, J.,Wurz, J., Shudes, C., Chang, C., & Shukla, M (2021, Oct 26). https://www2.deloitte.com/us/en/insights/industry/health-care/digital-transformation-in-healthcare.html

Vivek, ND (2022, Dec 22), Health Digitalization raises privacy concerns by on in; The Hindu Business line, https://www.thehindubusinessline.com/opinion/health-digitisation-raises-privacy-concerns/article33378752.ece <last visited on 24.08.2022>

Purdue Online.How has Technology changed Education. https://online.purdue.edu/blog/education/how-has-technology-changed-education#:~:text=Students%20can%20collaborate%20on%20group,roles%20of%20teachers%20and%20learners. <last visited on 19.8.2022>

Kapur, S. (2022, Aug 14). The Digitalisation of education: A boon or a bane; in Voices India; Times of India; https://timesofindia.indiatimes.com/blogs/voices/the-digitalisation-of-education-a-boon-or-a-bane/ <last visited on 19.8.2022>

Ainslee, J.,(2018, April 28). Digitization Of Education In The 21st Century. https://elearningindustry.com/digitization-of-education-21st-century <last visited on 19.8.2022>

Bhalla, A. (2022, June 2). Cash no longer the king: 65% transactions to be digital by 2026, says BCG-PhonePe Report; Financial Express https://www.financialexpress.com/industry/banking-finance/cash-no-longer-the-king-65-transactions-to-be-digital-by-2026-says-bcg-phonepe-report/2546559/<last visited on 21.08.2022>

❑❑❑

2

Systematic Review of Digital Business Model in Government Networks & Services

Laxmi

Assistant Professor, Department of Commerce, Rajdhani College, University of Delhi

Abstract

The goal of this chapter is to organise and summarise the research and conclusions regarding the use of digital business models to improve government networks and services. The systematic search process produced 30 research publications that were extracted. By reviewing the articles, findings and conclusions were made at the end which helps the future researchers in their future studies. The past two decades have seen a significant level of digital advancement in government, from the early e-government initiatives in the dot-com era to the launch of government digital service units and the adoption of commercial IT practices. Despite making progress, governments remained wired but not transformed. There is still too much emphasis placed on digitizing front-end services while ignoring or deferring more difficult issues such as reengineering the underlying operations, processes, and systems. This study examines how and whether the e-government initiatives of the central government in India have been reworked to be more user-friendly during the pandemic; or, if these models are flexible enough to incorporate economic changes.

Keywords: *financial inclusion, digital business model, Indian government, e-governance*

Introduction

One of the world's largest and fastest-growing digital markets is India, where more than half a billion people have internet access. However, the degree of adoption is inconsistent. With the advancement of technology and widespread connectivity, nearly every sector of India's economy is expected to undergo an unprecedented transformation. In addition to rationalising regulations, improving infrastructure, and even initiating Digital India, a program aimed at doubling the size of the country's digital economy, the Indian government has taken several steps to promote digital progress. However, much remains to be done for India to reach its full potential. As part of their efforts to drive digitisation, government agencies can collaborate with the private sector and place technology at the core of their operations to help drive the process. As a result, digital solutions can generate revenue for service providers, stimulate start-ups, and provide individuals with more reasons to go online, whether to register for a property sale or to utilise any other service provided by the government. Unfortunately, due to the pandemic, governments have been severely impacted, as very few were prepared to deal with the challenges created by such an emergency. The study's research aim are as follows:

- What are the aspects and ideas of progress in the digital business model in government networks and services?
- What are the discussions and findings for improving the digital business model in government networks and services?

Literature Review

Technology has been an effective strategy for the government to achieve such rapid growth. As internet penetration and smartphone penetration increase, the future of banking will be digital. Aadhaar-linked bank accounts enable financial inclusion to progress beyond traditional models through biometric identification. The government has pushed a differentiated banking model by licensing new financial institutions tiers through which mobile network operators and Fintech companies can offer banking services under the aegis of a Payment Bank license. Additionally, microfinance institutions (MFIs) are encouraged to utilize technology to remain competitive and to provide an incentive for them to become small finance banks. One particular factor contributing to the growth of digital payments was the government's abrupt demonetisation policy in November 2016 (Agarwal & Jain, 2019). As stated by Chatterjee and Das (2021), the

Digital India program has contributed to a technological rebellion in India as businesses are using smartphones and the internet to expand their business operations. As a result of technological advancements, our daily lives, our work, and our communication have been transformed significantly. A significant change in the financial sector has been the availability of banking services and the ability to conduct monetary transactions through mobile devices. Mobile financial services are expected to increase the number of people covered by the banking system. Introducing digital payments into the education, healthcare, and other social safety net sectors could result in significant personal benefits and enhance government and aid agency efficiency by reducing transaction expenses and leakage. Digital payments are predicted to become more convenient, secure, and affordable as mobile phone use increases and internet access becomes more widespread. Banking technology has advanced rapidly enough, and more importantly, it has been demonstrated that the poor can also now obtain loans (Kandpal & Mehrotra, 2019).

Based on the Reserve Bank of India's vision for 2020, more than 600 million banking accounts are expected to be opened and serviced by information technology in many channels. In many states, however, financial inclusion remains hindered by illiteracy, low-income savings, and the absence of bank branches. In order to expand reach, synergies between technology providers and banking channels are necessary. There should be synergies between core banking and micro-financial applications (Paramasivan & Ganesh, 2013). Similarly, Bansal (2014) notes that banks realised that the rural poor could also be worthy borrowers from banks, just as microfinance institutions in rural areas in India. In the rural areas, financial inclusion has become an integral part of banks' business, covering unbanked villages with populations over 2,000 and those below 2,000. Several innovative products and services have been introduced by financial institutions using cutting-edge technology (Erumban & Das, 2016). Further advancement in the Indian financial services industry is the availability of various payment options and the ease of access to services provided by digital platforms, such as mobile banking, bank cards, etc. (Priya et al., 2018). Additionally, mobile payment applications developed for smartphones have contributed to the development of mobile financial services (Laukkanen, 2016). Customers could participate in banking transactions at any time via their mobiles to make payments between individuals, pay bills for purchases made online, and perform other financial transactions (Chawla & Joshi, 2019). In their study, Rana et al. (2020) noted that digital financial services (DFS) may provide an array of financial services that are reliable, secure, and affordable to low-income consumers in developing countries through the application of innovative technologies, including the use of mobile phones, digital platforms, and e-money solutions. According to Balasubramanian and Kuppusamy (2020), women's access to advanced financial services such as debit and credit cards is impacted by financial literacy. For the analysis of the demographic factors that influence access to and utilisation of financial services and products, Dar and Ahmed (2020) used the World Bank index database. In this study, the determinants of financial inclusion and the barriers to inclusion were examined in detail. The study found that factors such as gender, age, educational level, and income are critical factors affecting debit card ownership and financial inclusion indicators, such as bank accounts, credit cards, and savings accounts. Innovations in technology and communication significantly impact the FI ecosystem's development, which in turn promotes a more equitable socio economic development. The level of transparency and efficiency of competition can be improved to reduce service delivery costs and strengthen administrative processes at the back end. In order to provide essential financial services, banks can design and develop customised products and ensure their effective implementation by leveraging FinTech and BC networks (Ozili, 2020; Murthy et al., 2019).

In the future, it will be increasingly important to empower consumers in relation to their internet activities. Indians, at over one billion people, are becoming increasingly data-rich due to the prevalence of mobile phones and internet connectivity, as it is in many other jurisdictions. Without public initiatives, however, newly created digital data is likely to be gathered and stored in private and proprietary data repositories. The challenge is resolving information inequalities and a lack of trust and enabling customers to quickly access and disclose their private information. Consumers will likely benefit from a more personalised product offering based on their individual needs with such an approach (Mhlanga, 2020). A common argument is that stand-alone meets the system's requirements without suffering delays, costs, or difficulties associated with overlay systems (such as correspondent banking networks). As a result of its interoperability, the Indian payment system illustrates how fast payment systems, based on public infrastructure, and overlay systems, are equally convenient for domestic transactions as stand-alone payment systems, in addition to the fact that the transactions are executed in financial terms within the boundaries of a regulated financial system. In contrast, stand-alone systems, such as stablecoins operated by large technology firms and others present risks that are yet to be adequately addressed (Siddik, Alam, & Kabiraj, 2020). There are also some cautionary lessons to be drawn from the rapid growth of India's financial infrastructure, which has not been able to be adequately examined in this report. It is common for digital financial platforms to rely on third-party agents, as was done with Aadhaar, given the importance of customer uptake. Fraudulent behaviour can result from this, and consumers who are less educated and technologically illiterate are at greater risk. Therefore, pre-emptive fraud detection and consumer education are necessary. Consumers must be continuously educated about the features and functions of digital financial platforms and how to safeguard private information. In addition, it is essential to note that not all jurisdictions can benefit from India's digital financial infrastructure advancement. Depending on the starting point, it may be appropriate to adopt a different model for different countries (Barik & Sharma, 2019).

Table 1: Annotated bibliography

Sl. No.	Title	Author (s)	Year	Country	Sample size	Mode of the study (offline/ online)	Findings
1	Business models in digital form in India: the government contribution and the road ahead	Reema and Susmi	2011	India	Literature survey	Online: Analytical approach/ secondary research	Identification of further initiatives that need to be taken by the Government of India.
2	Digitalization in India	Tanpreet and Sadhana	2017	India	Literature survey	Online: Secondary research	(1) Create awareness and ownership around digitalization. (2) Design a digitalization roadmap (3) Digitalize business model
3	E-Governance Initiatives in India with Special Reference to Punjab	Mahajan	2009	India, Punjab	Punjab government	Online: Secondary research	Through the use of Information Communication Technology (ICT), e-Government enhances relationships between government and government, government and citizens, citizens and government, and the government and the private sector.

Sl. No.	Title	Author (s)	Year	Country	Sample size	Mode of the study (offline/ online)	Findings
4	A review of e-government initiatives for rural citizens in India	Hemant	2008	Rural India	Rural India/E-govt growth model	Online: Secondary research	The public-private partnership model will greatly contribute to the proliferation of ICT networks across the country. In the event that this relationship is nurtured with due consideration for the future, the government can devote its resources and efforts to administration rather than developing and maintaining the physical infrastructure from conception to completion. The study agrees with the concept that the government should abandon efforts to 'reinvent the wheel.'
5	Impact Assessment of e-Governance Projects	Bhatnagar	2008	India	7,500 users and 13 RPOs/ POs, 9,500 users	Offline	The results of the study indicate the need to expand the coverage and accelerate the implementation of eGovernment.
6	Digital India: An Opportunity to Leapfrog in terms of Business and Technology	Harpreet, Prateek and Harshleen	2017	India	Random	Online and offline: Questionnaire and secondary research	It is generally considered that India presents a unique opportunity for DI projects, regardless of whether they are undertaken by the Government, the Private Sector or through Public-Private Partnerships. This will lead to inclusive growth in many fields, including manufacturing, electronic services, and devices, as well as tremendous employment opportunities for citizens.
7	Digital finance and inclusion in the time of COVID-19	Benni	2021	Rome	Global population	Online: Secondary research	Based on the past twenty years' experience, there is no doubt that the fintech sector continues to innovate and grow in the face of global crises, as new digital financial solutions are being introduced to fill the gaps found in traditional banking, thus meeting the need of crisis-affected populations for customized, more sophisticated, as well as accessible financial services.

Sl. No.	Title	Author (s)	Year	Country	Sample size	Mode of the study (offline/ online)	Findings
8	Digital Financial Services and the Pandemic: Opportunities and Risks for Emerging and Developing Economies	Agur, Soledad and Celine	2020	USA	Global population	Online: Secondary research	Digital financial services enable entrepreneurs to easily connect with banks, employees, suppliers, and new markets. They facilitate peer-to-peer transactions which also comprises remittances, and they make it easier for governments to reach households and businesses. Moreover, they can enable governments to reach households and businesses quickly. All of these factors become increasingly important during the COVID-19 crisis, as governments strive to ensure that funds are distributed quickly and efficiently to those in need, and many private households and businesses seek to access financial services online as soon as possible.
9	FinTech and Market Structure in the COVID-19 Pandemic	Financial Stability Board	2022	Global	Fintech services	Online: Secondary research	The study of market shares in retail financial services is limited by a lack of comprehensive data. However, market participants' insights and indirect evidence indicate that BigTechs have increased their presence in financial services.
10	India Digital Financial Inclusion: Journey Map Report	USAID	2019	India	Senior experts in govt, developmental and financial sector	Online and offline: Reports	A significant amount of this research has been devoted to the hypothesis that promoting digital payments will lead to growth within the DFS ecosystem and will, in turn, promote financial inclusion in poor and underserved populations. In light of the analysis, it is unclear whether electronic payments contribute to financial inclusion or alleviation of poverty.

Sl. No.	Title	Author (s)	Year	Country	Sample size	Mode of the study (offline/ online)	Findings
							Merchant payment digitization was deemed to be of limited significance in the context of financial inclusion. The linkages between digital payments and development need to be examined in greater detail in future programming.
11	A study to assess the impact of COVID-19 pandemic on digital financial services and digital financial inclusion in India	Mohd and Sana	2021	India	DFS	Online	The findings of this report indicate that there is a detrimental effect on financial inclusion as a result of COVID-19, whereas some DFS have been positively impacted.
12	Nexus between financial inclusion and economic growth: Evidence from the emerging Indian economy	Dipinsha	2016	India	Various dimensions of financial inclusion and economic development for the period 2004-2013	Online and offline: VAR models and Granger Causality test	Several studies have established a positive correlation between economic growth and financial inclusion, such as the penetration of the banking industry, accessibility to banking services, and deposit use.
13	Digital Financial Inclusion: A Payoff of Financial Technology and Digital Finance Uprising in India	Ravikumar	2019	India	Data from The World Bank, the Consultative Group to Assist the Poor (CGAP), National Payments Corporation of India (NPCI), Reserve Bank of India, and Alliance for Financial Inclusion (AFI).	Online: Descriptive study	The digital era of finance includes digital financial products and services. Financial technology companies have transformed the face of payments, credit, remittances, and insurances, and they are at the forefront of digital financial inclusion. Financial technology companies allowed adults in India to open bank accounts.

Sl. No.	Title	Author (s)	Year	Country	Sample size	Mode of the study (offline/ online)	Findings
14	Change in the uses pattern of digital banking services by Indian rural MSMEs during demonetization and Covid-19 pandemic-related restrictions	Shafique and Samiran	2021	India	148 rural MSME owners	Offline: IBM SPSS and AMOS	Convenience (including perceived ease of use and utility), self-efficacy perception, demonetization, perceived performance, and pandemic effect all contribute to attitudes toward the use of DBS.
15	Need for sustainable global business model in software outsourcing: The Indian perspective	Daisy and Reema	2013	India	Literature review search in the e-journals databases: ABI/ INFORM Complete, EBSCO Business Source Complete, Emerald Management Xtra and Science Direct.	Online: Secondary research	In order for India to become a dominant player in the outsourcing industry globally, sustainability is an important factor. In addition, the Indian software market is experiencing rapid growth, with an increasing number of businesses entering the sector due to the promising prospects. Therefore, both existing market players and new entrants require a model.
16	IndiGov-O: An ontology of Indian government to empower digital governance	Akshi and Arunima	2016	India	Semantic Web applications	Online	A government's stakeholders are its citizens and the government itself. In addition to facilitating communication between individuals, social media can provide a unique opportunity for citizens to be actively engaged in governance.
17	Twitter Based Sentiment Analysis of GST Implementation by Indian Government	Prabhsimran, Ravinder and Karanjeet	2018	India	IMDb and Demonetization datasets	Online: Semantic analysis	A majority of Indians supported the decision of the Indian government to implement GST.

Sl. No.	Title	Author (s)	Year	Country	Sample size	Mode of the study (offline/ online)	Findings
18	The design of digital financial infrastructure: lessons from India	Derryl, Frank and Siddharth	2019	India	e-KYC and other applications	Online: Secondary analysis	In India, the payment system is designed to challenge the viability of private payment systems by offering inexpensive and real-time payment facilities to citizens. The establishment of a legal framework for data fiduciaries will make it possible for individuals to easily gain access to the information generated through their online behavior and to determine under what conditions such information should be disclosed.
19	E-service quality model for Indian government portals: citizens' perspective	Debjani, Umesh and Gupta	2012	India	Focus group of experts consisting of practitioners, researchers and government officials	Offline: Exploratory factor analysis	A total of seven constructs were identified as a result of the analyses: "citizen centricity, transaction transparency, technical adequacy, usability, complete information, privacy and security, and usefulness of information". It is possible to use these principles to evaluate the quality of demand side service provided by public portals.
20	Role of digital economy and technology adoption for financial inclusion in India	Vishal and Priyanka	2021	India	433 educated adults (males and females) residing in different districts of Rajasthan (India)	Offline: Confirmatory factor analysis	Study participants were limited to adults who were educated and resided in a particular region within a country. Therefore, it is difficult to generalize results in terms of their relevance to other countries and regions.
21	Role of Digitization and E-commerce in Indian Economic Growth: An Employment Generation Perspective	Payel and Abhishek	2015	India	Employment opportunities	Online: Secondary analysis	India's GDP is increasing as a result of the impact of digitization. Aside from the obvious economic benefits of the Indian internet economy, the unquestionably successful model is also transforming the social lives of Indians, whose benefits are difficult to quantify.

Sl. No.	Title	Author (s)	Year	Country	Sample size	Mode of the study (offline/ online)	Findings
22	Connecting the unconnected: innovations in mobile government in the Indian state of Andhra Pradesh	Sundar	2020	India, Andhra Pradesh	Digital gram pratinidhi (DGP)	Online: Secondary analysis	DGPs recruited as DGPs may earn a higher salary because they handle a larger volume of electronic transactions in remote villages than their female counterparts. It has been suggested that, based on some limited evidence, the men DGP should have a college degree and should ideally be in their twenties. In addition to highlighting the constraints in ramping up these innovations, the study highlights some implications for the management of these innovations in the future.
23	A Study of the Readiness of Indian Banks to Absorb COVID-19's Impact Through New Emerging Technologies and Strategies for Competitive Advantage	Narinder and Kamal	2021	India	286 respondents who are bankers, financial planners, lenders, insurers, financial consultants, bank trainers, academicians and students studying post graduate courses in banking, finance and insurance companies	Offline: Purposive sampling technique	With higher sales resulting in higher consumption demand, the Indian economy has begun to show signs of recovery but still needs monitoring. Despite the moratorium being availed of by 40 percent of outstanding debt, the banking sector continues to deal with rising non-performing assets. A software solution based on artificial intelligence and machine learning, in conjunction with educating loan officers on current technology, could reduce NPAs. The future of a global economy lies in the implementation of the digital revolution and the development of new business models using Fintech.
24	Digital Transformation in the Age of COVID-19	OECD	2020	Universal/ global	Global data	Online: Secondary analysis	After reviewing current digital policies in the context of the COVID-19 emergency, governments will be presented with extensive, complex issues requiring international collaboration, cooperation, and dialogue.

Sl. No.	Title	Author (s)	Year	Coun-try	Sample size	Mode of the study (offline/ online)	Findings
							It is possible to inform decisions on a wide range of policy areas by combining the Digital Economy Outlook 2020 with the Going Digital Toolkit (OECD, 2020i). With its evidence-based analysis, policy advice, and development of international standards in areas such as artificial intelligence and data protection, the OECD will continue to support countries' digital transformations.
25	Digital Financial Inclusion in India: A Review	Gaurav	2019	India	Sec-ondary sources	Online: Secondary analysis	The new banking technology is designed to identify people's behavior in terms of adopting and using banking services.
26	Financial Inclusion landscape of India	ASSO-CHAM	2020	India	Sec-ondary sources	Online: Secondary analysis	While India has made great strides in financial inclusion, additional efforts from all stakeholders, including civil society, are required. The advent of digital connectivity, facilitated by focused efforts undertaken by all the responsible stakeholders, will be able to lead to full financial inclusion in the coming years. Several issues, such as tailor-made financial services for customers, increasing financial literacy, etc., need to be resolved to ensure that the efforts of government schemes are not wasted. With technology usage and infrastructure increasing, Fintech and Digital Financial Services will continue to grow in importance over the course of the future.
27	Digital Revolu-tion in the Indian Banking Sector	Shifa	2020	India	Sec-ondary sources	Online: Secondary analysis	Digitalization has assisted Indian banks in increasing their customer base as well as providing first-class services, as efficiency is taken into account. The smooth and long-term execution of this innovative business model depends on administrative compliance.

Sl. No.	Title	Author (s)	Year	Country	Sample size	Mode of the study (offline/ online)	Findings
							As we move into the digital age, one of the most important concerns is to ensure customer protection against online fraud and that the highest level of cybernetic values are utilized.
28	Digital transformation priorities of India's discrete manufacturing SMEs – a conceptual study in perspective of Industry 4.0	Gautam et.al.	2020	India	250 academic articles published until the end of September 2018	Online: Maturity assessment survey of Indian SMME	Several key insights can be gained from the maturity survey. For example, the self-assessment of the Indian SMME community indicates operational measures that are adopted before implementing production and design interventions in order to achieve the transformation they aspired to. As indicated by the survey, manufacturers intend to modify manufacturing and design strategies in accordance with key performance indicators; the first step would be to acquire real-time machine data, analyse the results, and integrate them into manufacturing and design decisions.
29	Digitalization and Indian economy: patterns and questions	Lazanyuk and Swati	2021	India	Secondary sources	Online: Secondary analysis	As a result of the ongoing pandemic crisis, consumers' habits are changing and the need for qualitative improvements in economic development approaches is escalating. If digitalization is implemented properly, the Indian economy may be able to become the third largest in the world in the future, in terms of economic growth.
30	Digital Payment and Its Discontents: Street Shops and the Indian Government's Push for Cashless Transactions	Joyojeet	2018	India	238 vendors	Offline: Data and theory-driven coding techniques	The use of digital payment systems rapidly expanded in the aftermath of the financial crisis, but declined following the introduction of new banknotes. When technology adoption challenges established practices, resistance occurs; these initiatives should be viewed in the context of the larger environment in which they are being applied, since the issue transcends the user experience and platform.

Discussions and Findings

Any country's financial sector is vital to its economic growth and development. Financial capability is a key factor in the stability and functioning of financial markets. An inclusive financial sector facilitates the accessibility and utilisation of basic economic resources. In recent years, financial inclusion through digital technology has contributed significantly to the development of the economy. In particular, the substantial correlation between digital financial inclusion and growth in the economy is not likely to be attributed to biases resulting from missing variables, contemporaneity, or reverse causality. Researchers have found that integrating digital financial services into payments can boost economic growth annually. There is a two-phase growth process in India's digital financial services sector. It is split into two phases: the first covers the period up to August 2016, the period of demonetisation, and the second covers the period after demonetisation. The volume of digital financial transactions grew steadily by two per cent per month. The introduction of numerous digital wallets and bank debit cards resulted in the growth of transactions through pre-paid instruments (PPIs) and debit cards at the beginning of the second phase. NPCI's (National Payments Council of India) digital payments platform will soon capture this space.

Future Implications

However, the benefits of financial inclusion do not come without new challenges. Many fintech companies rely on traditional banks to store customer funds. Insolvency of the bank hosting the cash balances of mobile money users could compromise these accounts. Customer funds may also be at risk if inadequate security is provided and/or if the service provider fails to handle customer funds in a prudent manner. In countries with low financial and digital literacy, this can undermine the inclusion of digital financial services. Increasingly, as technology is applied to a wide range of financial services, enhancing legal frameworks becomes vitally necessary to mitigate social and economic risks associated with stability. There is a requirement for legal frameworks to address possible concerns associated with Fintech, namely risks associated with cybersecurity, misuse of new payment channels for illegal purposes, such as financing of terrorism and money laundering, as well as risks associated with liquidity.

References

Bansal, S. (2014), "Perspective of technology in achieving financial inclusion in rural India", Procedia Economics and Finance, Vol. 11, pp. 472-480.

Paramasivan, C., and Ganeshkumar, V. (2013). "Overview of financial inclusion in India". International Journal of Management and Development Studies, Vol. 2, No. 3, pp. 45-49.

Chatterjee, A. and Anand, N. (2017), "Financial Inclusion, Information and Communication Technology Diffusion and Economic Growth: A Panel Data Analysis", working paper 165/ 2017, Madras School of Economics, Chennai, India, July 2017.

Erumban, A. A. and Das, D. K. (2016), Information and communication technology and economic growth in India, Telecommunications Policy, 40(5): 412-431.https://doi.org/10.1016/j.telpol.2015.08.006

Priya, R., Gandhi, A.V. and Shaikh, A. (2018), Mobile banking adoption in an emerging economy: An empirical analysis of young Indian consumers, Benchmarking: An International Journal, 25(2): 743-762. https://doi.org/10.1108/BIJ-01-2016-0009

Laukkanen, T. (2016), Consumer adoption versus rejection decisions in seemingly similar service innovations: The case of the internet and mobile banking, Journal of Business Research, 69(7): 2432-2439. https://doi.org/10.1016/j.jbusres.2016.01.013

Chawla, D. and Joshi, H. (2019), Consumer attitude and intention to adopt mobile wallet in India–An empirical study, International Journal of Bank Marketing, 37(7): 1590-1618. https://doi.org/10.1108/IJBM-09-2018-0256

Rana, N.P. Luthra, S., & Rao, H.R. (2020), Key challenges to digital financial services in emerging economies: the Indian context, Information Technology & People, 33(1): 198-229. https://doi.org/10.1108/ITP-05-2018-0243

Balasubramanian, S.A. and Kuppusamy, T. (2020), Does female labour force participation contribute to better financial inclusion? Evidence from cross-country analysis, Journal of Economic and Administrative Sciences, ahead-of-print of print. https://doi.org/10.1108/JEAS-08-2020-0151

Dar, A. B., & Ahmed, F. (2020). Financial inclusion determinants and impediments in India: insights from the global financial inclusion index. Journal of Financial Economic Policy.Ozili PK (2020) Financial inclusion research around the world: a review. Forum for Social Economics

Murthy, G., Fernandez-Vidal, M., Faz, X., & Barreto, R. (2019). Fintechs and financial inclusion: looking past the hype and exploring their potential. Fintechs and financial inclusion: looking past the hype and exploring their potential.Agrawal, G., & Jain, P. (2019). Digital financial inclusion in India: a review. Behavioral Finance and Decision-Making Models, 195-203.

Kandpal, V., & Mehrotra, R. (2019). Financial inclusion: The role of Fintech and digital financial services in India. Indian Journal of Economics & Business, 19(1), 85-93.

Mhlanga, D. (2020). Industry 4.0 in finance: the impact of artificial intelligence (ai) on digital financial inclusion. International Journal of Financial Studies, 8(3), 45.

Siddik, M., Alam, N., & Kabiraj, S. (2020). Digital finance for financial inclusion and inclusive growth. In Digital transformation in business and society (pp. 155-168). Palgrave Macmillan, Cham.

Barik, R., & Sharma, P. (2019). Analysing the progress and prospects of financial inclusion in India. Journal of Public Affairs, 19(4), e1948.

Agrawal, G., & Jain, P. (2019). Digital financial inclusion in India: a review. Behavioral Finance and Decision-Making Models, 195-203.

Chatterjee, A., & Das, S. (2021). Information communication technology diffusion and financial inclusion: an inter-state analysis for India. Innovation and Development, 11(1), 1-23.

Ozili, P. K. (2020). Theories of financial inclusion. Uncertainty and challenges in contemporary economic behaviour. Emerald Publishing Limited.

Annotated Bibliography References

Khurana, R., & Routray, S. (2011). Business models in digital form in India: the government contribution and the road ahead. *International Journal of Indian Culture and Business Management*, *4*(5), 572-588.

Narula, T. & Rana, S. (2017). Digitalisation in India. *International Journal on Emerging Technologies (Special Issue NCETST-2017) 8(1): 298-303(2017)*

Mahajan, P. (2009). E-Governance initiatives in India with special reference to Punjab. *Asia-pacific journal of social sciences*, *1*(1), 142-155.

Patel, H. (2008). A Review of E-Government Initiatives for Rural Citizens in India. *8(3): 3-4*

Bhatnagar, S. C. (2008). Impact assessment of e-governance projects: A benchmark for the future.

Jaswal, H., Bajaj, P. & Kaur, H. (2017). Digital India: An Opportunity to Leapfrog in terms of Business and Technology. *International Journal Of Engineering Research & Technology (IJERT).*

Benni, N. (2021). Digital finance and inclusion in the time of COVID-19: Lessons, experiences and proposals. *Food and Agriculture Organization of the United Nations.*

Agur, I., Peria, S. M., & Rochon, C. (2020). Digital financial services and the pandemic: Opportunities and risks for emerging and developing economies. *International Monetary Fund Special Series on COVID-19, Transactions, 1, 2-1.*

Financial Stability Board. (2022). FinTech and Market Structure in the COVID-19 Pandemic: Implications for financial stability.

USAID. (2019). India Digital Financial Inclusion: Journey Map Report. Retrieved from, https://www.usaid.gov/digital-development/india-digital-financial-inclusion-journey-map-report

Shafeeq, M., & Beg, S. (2021). A study to assess the impact of COVID-19 pandemic on digital financial services and digital financial inclusion in India. *African Journal of Accounting, Auditing and Finance*, 7(4), 326-345.

Sharma, D. (2016). Nexus between financial inclusion and economic growth: Evidence from the emerging Indian economy. *Journal of financial economic policy.*

Ravikumar, T. (2019). Digital financial inclusion: A payoff of financial technology and digital finance uprising in India. *International Journal of Scientific & Technology Research*, 8(11), 3434-3438.

Ahmed, S., & Sur, S. (2021). Change in the uses pattern of digital banking services by Indian rural MSMEs during demonetisation and Covid-19 pandemic-related restrictions. *Vilakshan-XIMB Journal of Management.*

Jain, D. M., & Khurana, R. (2013). Need for sustainable global business model in software outsourcing: the Indian perspective. *Business Process Management Journal.*

Kumar, A., & Joshi, A. (2016, August). indigo-O: an ontology of Indian government to empower digital governance. In *2016 1st India International Conference on Information Processing (IICIP)* (pp. 1-6). IEEE.

Singh, P., Sawhney, R. S., & Kahlon, K. S. (2019). Twitter-based sentiment analysis of GST implementation by Indian government. In *Digital Business* (pp. 409-427). Springer, Cham.

D'Silva, D., Filková, Z., Packer, F., & Tiwari, S. (2019). The design of digital financial infrastructure: lessons from India. *BIS Paper*, (106).

Bhattacharya, D., Gulla, U., & Gupta, M. P. (2012). E-service quality model for Indian government portals: citizens' perspective. *Journal of enterprise information management*, *25*(3), 246-271.

Vyas, V., & Jain, P. (2021). Role of digital economy and technology adoption for financial inclusion in India. *Indian Growth and Development Review*, *14*(3), 302-324.

Chaudhuri, P., & Kumar, A. (2015). Role of digitisation and e-commerce in Indian economic growth: an employment generation perspective. *Research Gate. December.*

Balakrishna, S. (2020, September). Connecting the unconnected: innovations in mobile government in the Indian state of Andhra Pradesh. In *Proceedings of the 13th International Conference on Theory and Practice of Electronic Governance* (pp. 442-451).

Bhasin, N. K., & Gulati, K. (2021). A Study of the Readiness of Indian Banks to Absorb COVID-19s Impact Through New Emerging Technologies and Strategies for Competitive Advantage. *E-Collaboration Technologies and Strategies for Competitive Advantage Amid Challenging Times*, 50-75.

OECD. (2020). Digital Transformation in the Age of COVID-19: Building Resilience and Bridging Divides, Digital Economy Outlook 2020 Supplement, OECD, Paris, www.oecd.org/digital/digital-economy-outlook-covid.pdf.

Agrawal, G., & Jain, P. (2019). Digital financial inclusion in India: a review. *Behavioral Finance and Decision-Making Models*, 195-203.

ASSOCHAM. (2020). Financial Inclusion landscape of India.

Fathima, J. (2020). Digital Revolution in the Indian Banking Sector. *International Journal of Commerce, 8*(1), 56-64.

Dutta, G., Kumar, R., Sindhwani, R., & Singh, R. K. (2020). Digital transformation priorities of India's discrete manufacturing SMEs–a conceptual study in perspective of Industry 4.0. *Competitiveness Review: An International Business Journal.*

Lazanyuk, I., & Modi, S. (2021). Digitalisation and Indian economy: patterns and questions. In *SHS Web of Conferences* (Vol. 114). EDP Sciences.

Pal, J., Chandra, P., Kameswaran, V., Parameshwar, A., Joshi, S., & Johri, A. (2018, April). Digital payment and its discontents: Street shops and the Indian government's push for cashless transactions. In *Proceedings of the 2018 CHI Conference on Human Factors in Computing Systems* (pp. 1-13).

3

Digital Mindset

Dr Ajay Bhatia
Visiting Faculty, IBMR Business School, Gurgaon

Abstract

Technology has been disrupting almost all walks of our work and life. Digital forces, like big data, artificial intelligence, robotics, mobile, 3-D printing, virtual reality, cloud computing, Internet of things (IOT), and social media, etc, coupled with Covid-19 pandemic, have all hastened the pace of change happening all around us. While recent trends show that the majority of technology spending has been going towards digital transformation, however research shows that more than 70% of such transformations are failing for want of the right digital mindset.

Digital mindset is less about technology and more about reimagining the future by adopting a new set of digital habits, attitudes, capabilities, behaviour, customer centricity, and bringing together people and technology, that demonstrates autonomy, openness, innovation and speed. It allows you to choose to think differently about the complete process of organizational change and development while unleashing the potential of digital transformation, be it collaboration, computation, democratizing teams and processes, building efficiencies, making business scalable and changing organizational gears to manage disruptions with composure.

The current chapter has endeavoured to break the myths around digital mind-set, clarify why it is relevant and virtually a "make or break" condition for any organization envisioning their future in this digital age. The chapter also details out well-researched key principles and common threads that constitute a digital mindset, and finally how should an organization go about systematically fostering a digital mindset.

Keywords: *Digital mindset, digital transformation, digitalisation, digital culture, growth mindset.*

Introduction

The evolution of our industrial environment since the last century, and the way it has been redefining the personal and professional lives of people, has gone through several distinct phases of "Industrial Revolution". The "1st industrial revolution" that took place at the end of 18th century, was marked by mechanisation, and enabled by steam and water power. By the start of the 20th century, the "2nd Industrial revolution" dawned in, which was powered by electricity and characterised by mass, assembly-line production. Later, the 1970s saw the onset of the "3rd Industrial revolution" often called "digital revolution", involving the development of computers, electronics and IT (information technology) to further automate machines and production processes. The "4th Industrial Revolution" is the currently developing environment, which is primarily characterized by amalgamation of technologies, be it Internet of Things (IoT), robotics, virtual reality (VR) or artificial intelligence (AI), etc., that is gradually obscuring the demarcation between the physical, digital, and biological domains. It is unlike any previous industrial revolution whether in terms of scale, scope, or complexity, being primarily characterised by explosiveness of its development and the disruptiveness of its technologies. It is fundamentally altering the way we live, work, and relate to one another.

The Covid-19 pandemic has hastened the onset of remote/ hybrid work era, wherein people have been learning to work though machines and using diverse digital technologies to stay connected and perform their work while being aligned to shared goals, norms and new organizational culture.

The digital wave has moved well beyond the industries. Digitalisation is impacting almost all walks of our work and life, virtually requiring us to reimagine our daily realities, be it getting education, entertainment, buying groceries, local transportation, buying insurance, transferring money through mobiles, searching for vendors, recruitment, training, engagement, marketing, operations, supply chain, logistics, etc. While technologies got introduced in the previous era as well, but what now becomes the differentiator are the people behind the technology and how do the two integrate with each other to achieve common goals and deliver greater value to the business.

"Digitalisation happens at the intersection of technology and people"

For an organization to succeed, it must have the right people before anything else, who in turn need to have the right mindset in order to fully exploit the potential of digitalisation. With the rapid proliferation of digital technologies and its adoption also increasing exponentially, especially post Covid-19 pandemic induced virtual working, many people have got used to digital tools like Zoom, Google Meet, MS-Team, Google drive, Google sheets, intranet, online systems, etc., to work seamlessly across physical or geographic boundaries. The shift from analog to digital mindset is underway!

It's important here to understand that "Digitisation" (i.e., the process of turning physical data into digital data), led us to "Digitalisation", which makes use of digitized information to improve systems and processes. "Digital transformation", which is taken up by any digital enterprise or business, is the outcome of digitalisation. A digital business uses "digital technologies and resources to drive day-to-day business operations, develop revenue and performance, optimise both external and internal processes and business activities, and deliver personalised, meaningful experiences to customers" (McKendrick, 2020). It is changing the way organisations use and think about technology, moving technology from a supporting player to a leading player in innovation, revenue and market growth. As per a Gartner study authored by Stamford (2014), a "digital business cannot be regarded as an IT program but instead needs to become an enterprise mindset and lingua franca, with digital expertise spread across the enterprise and value ecosystem".

Digital transformation is reorienting how an organization uses technology, people, and processes in their quest of new business models and innovative revenue streams, to manage the changing customer expectations across products and services. It generally entails the integration of digital technology into all areas of a business resulting in fundamental changes to how businesses operate and how they deliver value to customers. As the world around us is increasingly becoming digital, it has become imperative for companies to embark on a digital transformation journey to remain competitive and relevant in today's dynamic business environment.

Vail (2019) research underlined that "along with more information, computing, communication, and connectivity, digital technologies also enable new forms of collaboration among distributed networks of diversified actors". This eventually creates critical dependencies among its diverse stakeholders, whose interests may not necessarily be fully aligned, and allows organizations to explore their potential to innovate, achieve high standards of performance and make a difference to the people, industries, and the society.

Goes without saying that such a digital transformation is not possible without radical changes in culture, as it requires organizations to continually challenge the status quo, take risks, and develop tolerance for ambiguity and failure (Reinhard et al., 2016). This sometimes means shifting away from age-old business processes that companies had built and developed, to be replaced by relatively new and evolving practices.

"It's no longer enough to get up to speed with the latest technology. The way of working and thinking also needs to change. Digitalization offers new rules of the game and opportunities. Those who understand the new environment will win the race."- Nick Jue, CEO of ING Germany

While data is important in the digital age, however, more important is to understand what data you will collect, what do you intend to do with it, what skills and technologies are required to handle it, how will you store it and keep safe, how will you mitigate the risks, how do you interpret and analyze to gain insights that drive organizational value. As per (Gartner, 2014), first step in building a successful digital enterprise is creating the right digital mindset and shared understanding.

"A digital mindset requires us to fully understand how to think about data, how to analyze data, and how to ask all of the right questions to ensure that no harms or risks are embedded in them as well." (Neeley et al., 2022)

Decoding Digital Mind-Set

Psychologists describe "mindset" as "a way of thinking and orienting to the world that shapes how we perceive, feel, and act." It refers to "personality traits that can be changed in the long term" and thus "lies between attitude (i.e., your pre-conceived notion, belief or stance on a particular topic) and relatively more rigid personality traits".

Albert Einstein said, "The world as we have created it is a process of our thinking. It cannot be changed without changing our thinking."

A digital mind-set is also a way of thinking, but about data, devices, and technologies as well as how they operate in a digital organization, in conjunction with other resources and stakeholders.

A "digital mindset" can be defined as the "personality traits or behavioural dispositions that are not directly observable and significantly structure and channelize thinking, feeling, and acting in the social and operational contexts of digital transformation. The expression of personality dispositions does not produce an absolute, but probable readiness to behave, which can be moderated by external environmental factors, such as framework conditions, or new learning experiences." (Julian Knorr, 2022)

In comprehensive research on the topic by Solberg, Traavik, and Wong (2020, p. 17), a digital mindset was defined as "employees' individually held, general beliefs regarding personal and situational resources in the context of technological change".

In his Book- *"The Future Leader: 9 Skills and Mindsets to Succeed in the Next Decade"*, Jacob Morgan, a futurist and a globally recognized thought leader, after interviewing 140 top CEOs from around the world in 2020, highlighted the nature of digital mindset that future leaders need to have, which includes the mindsets of *"Global citizen"* (i.e., "thinking globally and embracing diversity), *"Servant"* (i.e., Practice humility and mindset to serve others), *"Chef"*(i.e., embracing technology and balancing it with humanity to use it for improving efficiency in the organization while also providing a sense of purpose and caring for employees) and *"Explorer"* (i.e. curiosity and openness to new ideas), besides the skills as of a *"Coach"* (motivate, inspire, engage and care about their teams), *"Futurist"* (i.e., staying on top of the future trends and leading the change), *"Technology Teenager"* (i.e., tech savvy and digitally fluent, with mindset to embrace technology and leverage it to best serve their company), *"Translator"* (i.e. communicating and connecting with people), and a *"Yoda"* (i.e. emotionally intelligent) (Morgan, 2020).

A digital mindset is less about the "digital skills" like programming languages or software knowledge, but rather more about realizing one's potential by leveraging the skills, competencies, resources, and technologies, in alignment with the changes in the business environment. Such a mindset is best summarized as *"personality traits that are relevant for success in digital transformation, and which can be changed in the long term, as per evolving needs"*.

Tsedal Neeley and Paul Leonardi from Harvard Business School, talked about a "30 percent rule" in their new book, "The Digital Mindset: What It Really Takes to Thrive in the Age of Data, Algorithms, and AI". As per this rule, if 100% stands for attaining mastery in a new skill, then any person just needs 30% to attain fluency in that skill (Harvard Business Review Press, May 2022). Same rule when applied to organisations, means that you don't need to be an IT-pro and master in all digital technologies in order to metamorphose into a digital organization. With approx. 30% of digital literacy, coupled with digital mindset, which includes inter-alia, how you think about using specific skills and technologies, how willing and risk-taking you are, how much you think your organization can afford to invest and change, etc., one should be in a good position to lead a digital organization to success.

Unlike typical change management programme which run for a finite period and have a defined beginning and end, once one adopts a digital mindset then there is no finite end-point. Organizations permanently remain in the state of transition, either by adopting/ adapting to emerging digital tools and technologies, or fine-tuning the organizational structure, systems and culture to align with the ongoing change, or leaders always keep working on the unfinished agenda while parallelly developing new agenda as per the evolving landscape.

So, digital mind-set is primarily about *"continuous learning, improvement and adaption"*. People should always remain curious, and keep developing their capacity to learn, unlearn and relearn faster. They need to adapt their styles, learn new skills and change their attitudes towards work, people and processes, as per evolving business expectations and potential opportunities on the horizon.

Why Digital Mindset is Relevant?

It has been reiterated in a recent study that "digital transformation is not just a traditional IT back-end process; rather it affects the organization as a whole, redefining strategies, entrepreneurial processes, innovation, governance mechanisms, and how a firm organizes its value chains and relationship networks that are so prevalent in our digital ecosystems and marketplaces." (Cennamo et al., 2020)

Learning new technological skills is essential for digital transformation, but not enough. Employees need to be "motivated to use their skills to explore potential", and "demonstrate a set of attitudes and behaviours that enable people and organizations to see how data, algorithms, and AI open up new possibilities and to chart a path for success in an increasingly technology-intensive world". (Neeley et al., 2022)

According to marketing intelligence firm IDC, "40 percent of all technology spending in 2021 was going towards digital transformation, with companies investing more than $2 trillion into the process". But Forbes estimated that "70 percent of digital transformation initiatives will fail, leaving a $900 billion hole in enterprise strategy". The simple reason attributed to this delta is the prerequisite of a growth mindset for any digital transformation to be successful. Unfortunately, most organizations prefer technology before investing in people.

"In the digital age, choosing not to develop a digital mindset leads to catastrophic repercussions."

One of the key components of Digital mindset is the *"growth mindset"*, a term popularised by Microsoft, however, this nomenclature was done by Stanford professor Dr Carol Dweck, who classified people's mindsets broadly into two categories, viz., *"Fixed Mindset"*, which believes that "intelligence and talent are static", mistakes are penalized, creativity is discounted on the altar of seniority, which pulls the organization downwards. In contrast, with *"Growth Mindset"*, one believes that "anyone can develop their skills with motivation, efforts and right leadership". The companies embracing "growth mindset" were seen to be "encouraging employees to experiment", "learn through failure", and "constantly push themselves to do better". With such an approach and philosophy, the "Growth mindset" companies quickly outpace 'fixed mindset' ones, to become more adaptive and successful in future.

In the new world order, an apparent digital divide is emerging. The perceived threat to professionals in the digital era is not from their job getting replaced by machines, algorithms and robots, but from those human beings who have imbibed relevant digital skills and mind-set. Employees' beliefs about technological change (viz., their "digital mindsets") are likely to influence their engagement in, or withdrawal from their company's digital transformation initiatives. "Employees' beliefs regarding the malleability of personal ability (fixed/growth mindset) and their beliefs about the availability of situational resources (zero-sum/expandable-sum mindset) influence the extent to which they see new technologies as providing opportunities for professional growth or as encroaching on their ability to display competency." (Solberg et al., 2020)

Employees with a digital mindset are likely to be more successful in their jobs, garner more rewards and challenging assignments, be happier and have higher satisfaction at work, with no dearth of alternate job opportunities elsewhere. Such people, irrespective of years of experience, qualifications or position held, are destined to take the centre stage, while those without digital mindset will gradually go into oblivion.

Similarly, leaders of today's modern organization have to think differently. They not only need to have the knowledge of the basic tenets of new technology, like coding, programming, algorithms, artificial intelligence, machine learning, robotics and 3-D printing, etc., but also understand the biases and grey side of new technologies, and most importantly, develop the capability to ascertain the integrity of the data presented and deriving accurate insights aligned to the desired behavioural and cultural change that the organization wants to bring about. Leaders who have a digital mindset are better able to build an adaptable, agile and resilient organization that is able to take advantage of the emerging opportunities and remain successful even during VUCA (volatile, uncertain, complex and ambiguous) times.

Principles of Digital Mind-Set

There is no hard & fast recipe of the right "digital mindset", since it is multi-dimensional. It manifests itself differently across different industries, organizations, and roles, each of which have certain strengths and characteristics essential to understand its essence.

As per KPMG research, a "digital mindset", across any individual, team, or an organization, hinges on the fine balance of four principles:

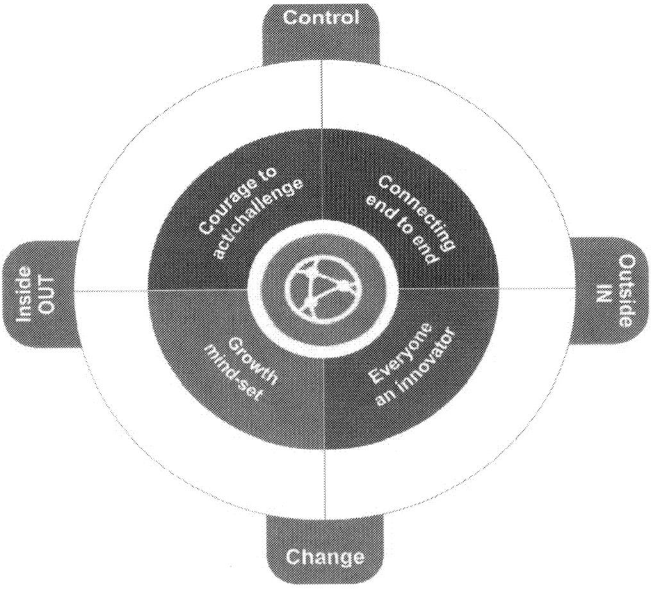

Source: https://home.kpmg/xx/en/home/insights/2020/07/kpmg-digital-mindset.html

1) **"Courage to act/challenge":** Organization focuses on extensive organization-wide communication (on goals, priorities and strategies) while building trust, transparency and interdependencies across departments and functions. Individual employees are given freedom to act while holding accountability for meeting business objectives. Leadership is not attached to a managerial position, as everyone is encouraged to act and challenge.
2) **"Connecting end to end"**: A digital mind-set sees the organization as a whole, applies design thinking, and connects its various departments, functions, products, services, people and technology to offer superior value, business outcomes, and customer experience
3) **"Growth mindset":** It involves the institutionalising of an innovative and flexible organizational culture that encourages agility, collaboration across role boundaries, learning orientation and change both in the present and futuristic context.
4) **"Everyone an innovator":** An organization is able to achieve its true potential only by truly believing that ability to innovate is not the domain of the identified few. Everyone can productively innovate how their and department's tasks, roles, and operations are coordinated through balanced integration of humans and machines.

The other common threads ([2]) on having a digital mindset are elaborated below:

- *No siloed approach:* An organization or a leader with a digital mindset never acts in a siloed manner, involves all stakeholders in charting out the new journey, focuses both on seeking "buy-in" and enhancing others' "capacity to learn", takes pains to "enlist voluntary support and participation from employees", "makes efforts to evolve organizational systems and processes", and has "bias towards radical and agile approach towards digital transformation". For any such digital transformation initiative to be successful, companies tend to strike a fine balance between enablement of digital organizational culture and institutionalization of enabling systems and processes.
- *"Agile methodologies",* as a set of principles, aligns with a digital mindset, as it makes an organization more adaptable and responsive to the needs, expectations and feedback of its customers and clients. A digital programme that does

not adopt agile principles, will always struggle to deliver the requisite digital transformation.

- *Leadership* is in a perpetual state of inventing, reinventing, and transitioning, and show increasing confidence to leverage data insights for affective decision making. Leaders ensure that their organizations develop a digital mindset while being capable of responding to the disruptions caused by the digital technologies (Benlian and Haffke, 2016).
- *Customer centricity:* A digital mindset come with a strong belief that "customer always comes first". Whether it is dynamically gathering and acting on customer feedback, or adding features that are beneficial to the customer, or doing quick turnaround on customer issues, or adopting lean methodologies to reduce wastage and lower costs for the customers.
- *Proactive approach:* Digital mindset envisions the new future, enables changes by galvanising people, technology and all other resources to meet the goals of digital transformation.
- *Pervasive:* Weaves a program that is at the heart of the organization, evident across the workforce, and involves everybody to succeed in the rapidly changing times.
- *Driver for change: A* digital mindset ensures that the desired service to be offered to the customer defines the business processes, and is not defined by them. The offline processes should not be blindly replicated in the online mode.
- *User Research:* A digital mindset lays undue focus on user research, to understand user requirements and expectations seamlessly and continuously, and makes it easier for them to use and consume organizations products and/or services, which ultimately leads to long-term organization success.
- *Openness:* Organization demonstrates openness to explore new ideas, different tools and systems. No single digital tools, or technology or solution is treated as a panacea for all the issues which an organization is confronted with.
- *Appetite for risk taking:* The mindset allows one to fail while experimenting with new and innovative idea, or "challenging the status quo".

How to Foster Digital Mind-Set

A 2015 study of Capgemini Consulting pointed out four key factors that play a critical role in determining the company's capability to be a digital organization, viz., "digital-first mindset", "digitized practices", "empowered talent", "data access" and "collaboration tools" (Bonnet et al., 2015). The study further elaborated that in order to build a "digital mindset", leaders must "explain the benefits of the digital transformation" to key stakeholders, "be a role model" to employees, and then to "introduce monetary as well as non-monetary rewards" to facilitate "digital change".

Research, expert views and past organizational experience of the author, point towards a few fundamental ways to successfully foster a digital mindset[1]:

- Create a sense of urgency
- Define and communicate organizational digital vision, goals and strategy
- Establish an empowered team of influential leaders to drive the change throughout the organization
- Empowered team does extensive communication and ensures wider participation of employees in both identifying the road map, laying down organization-wide milestones and team/ individual accountability
- Identification of individual strengths, weaknesses and talent gaps, in terms of the business objectives.
- Align individual work-related goals, personal development goals to orient individual employees achieve identified milestones
- Earmark training and competency-enhancement resources to bridge employees' capability gaps. It could include annual training budget, a department/business-unit/ function level upskilling program monitored by the respective heads, or access to e-learning modules, or sponsoring for specialized external training for a few followed by "Train the Trainer" approach for organization-wide rollout, networking opportunities, mentoring and coaching,
- Institutionalize a culture characterised by listening, collaboration, innovation, curiosity, transparency, patience, care, compassion and customer centricity.
- It's widely understood that *"Culture eats strategy for breakfast."* For any digital transformation initiative to succeed, leadership has to remember that *"Change happens at the speed of culture."*
- A 2017 research study on "Digital culture" had pointed out the key differences between an Analog and a Digital culture (Wokurka et al.,2017)

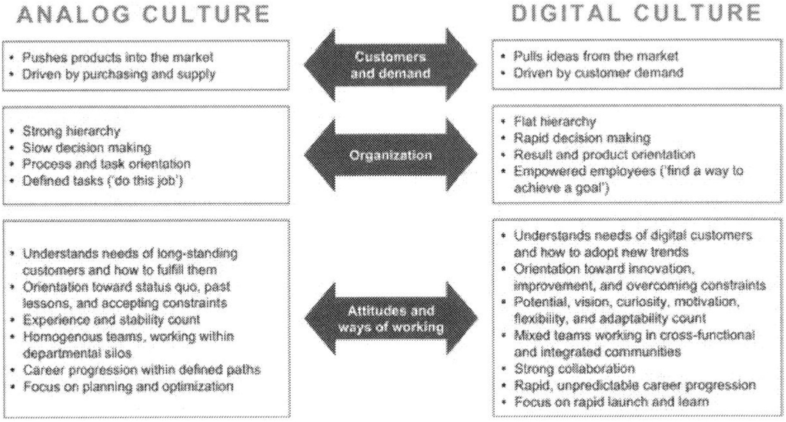

Source: https://doi.org/10.1007/978-3-319-40967-2_5

Some of the key points above were reinforced in other research studies as well which found the "need for firms to cultivate a willingness to take risks and to experiment with digital technologies using an agile approach for scalable transformation" (Fehér and Varga 2017; Dremel et al. 2017). Kontić & Vidicki (2018) stressed that "proactive leadership and "investment" are the critical aspects that influence a company's potential to become a "digital organization".

A digital culture is primarily enabled through leadership actions ("walking the talk"); laying down facilitating systems, policies and processes; robust management control over the entire program; total reward & recognition system; and leadership team always demonstrating growth mindset in their thoughts and actions. It is very important to keep employees engaged, measure and monitor their experience and keep supporting them in their journey. A Gallup study found that "engaged employees outperform their colleagues by 147 per cent".

The digital disruption caused by Uber in the taxi business was both "Technological (GPS, Mobile internet, apps, real time data) and Business-model related (frictionless, moment of need, ownership free, aggregated need, aggregated supply, real time pricing)" (Knight, 2018).

Google's famous *"20-Percent Rule"*, which has been the harbinger of its diverse innovations, is a classic example of institutionalizing such a culture, where employees can comfortably spend "20 per cent of their work time on personal projects" that individual employee is passionate about that will also help Google in realizing its vision and mission.

Microsoft also popularized the catch-word- *"Mobile-First, Cloud-First"* to align individual employees thinking and action required for digital transformation. Similarly, their campaign of *"One Microsoft"*, influenced its employees to demonstrate boundary-less behaviour, bring agility in innovation and improve customer experience. Similarly, Amazon's growth-mindset driven culture primarily fuelled its digital transformation journey, and enabled it to quickly progress from being an online book seller, to an all-encompassing e-commerce platform by continuously evolving its digital infrastructure to offer diverse products and services like "Amazon Prime Video", "Amazon Music", "Appstore for Android", "Echo" and "Alexa", "Fire Tablet" and TV, "Kindle" e-readers and books, "Amazon Web Services" (AWS), merchant products, vendor products, and almost everything under the sun.

Conclusion

Digitalisation and digital transformation, whether by individuals or organizations, are imperative to succeed in today's environment, characterised by rapid technological advancements, cut throat competition and disruptions. *"A journey well begun is a journey half done"*. The most meaningful starting point and something which sustains an organization throughout its transformation journey, is the digital mindset. With the right mindset, you are able to redesign customer and employee experience, redefine processes and systems by leveraging right technology as an enabler, build an agile and connected enterprise with a future-ready workforce, demonstrate boundary-less and customer-centric behaviour, and achieve sustainable success in your digital transformation journey.

References

Benlian, A., & Haffke, I. (2016). Does mutuality matter? Examining the bilateral nature and effects of CEO–CIO mutual understanding. The Journal of Strategic Information Systems, 25(2), 104-126.

Bonnet, D., Puram, A. D., Buvat, J., KVJ, S., & Khadikar, A. (2015). Organizing for digital: why digital dexterity matters. Capgemini Consulting, 16.

Cennamo, C., Dagnino, G. B., Di Minin, A., & Lanzolla, G. (2020). Managing digital transformation: scope of transformation and modalities of value Co-generation and delivery. *California Management Review*, *62*(4), 5-16.

Knight, W. (2018). Disruption – The Uber story. Think Digital First. https://thinkdigitalfirst.com/2018/01/31/disruption-uber-story/. Accessed on July 25, 2022

Kontić, L., & Vidicki, Đ. (2018). Strategy for digital organization: Testing a measurement tool for digital transformation. Strategic Management, 23(1), 29-35.

McKendrick, J. (2020). Digital is great, but where are the new business models? Forbes. https://www.forbes.com/sites/joemckendrick/2020/07/10/digital-is-great-but-where-are-the-new-business-models/?sh=777694755f30. Accessed on July 25, 2022

Morgan, J. (2020). The Future Leader: 9 Skills and Mindsets to Succeed in the Next Decade. John Wiley & Sons.

Neeley, T., Leonardi, P. (2022). Developing a Digital Mindset: How to lead your organization into the age of data, algorithms, and AI. Harvard Business Review Press. https://hbr.org/2022/05/developing-a-digital-mindset. Accessed on July 25, 2022

Reinhard, G., Jesper, V., & Stefan, S. (2016). Industry 4.0: Building the digital enterprise. *2016 Global Industry 4.0 Survey*, *1*(1), 1-39.

Solberg, E., Traavik, L. E., & Wong, S. I. (2020). Digital mindsets: Recognizing and leveraging individual beliefs for digital transformation. *California Management Review*, *62*(4), 105-124.

Stamford, C. (2014). Gartner identifies six key steps to build a successful digital business. https://www.gartner.com/en/newsroom/press-releases/2014-05-21-gartner-identifies-six-key-steps-to-build-a-successful-digital-business. Accessed on July 25, 2022

Vail, G. (2019). Understanding digital transformation: A review and a research agenda. The Journal of Strategic Information Systems, 28(2), 118-144

Wokurka, G., Banschbach, Y., Houlder, D., Jolly, R. (2017). Digital Culture: Why Strategy and Culture Should Eat Breakfast Together. In: Oswald, G., Kleinemeier, M. (eds) Shaping the Digital Enterprise. Springer, Cham. https://doi.org/10.1007/978-3-319-40967-2_5

Endnotes

[1]https://online.rmit.edu.au/blog/how-develop-digital-growth-mindset-your-team

[2] https://www.infoq.com/articles/meaning-being-digital/

4

Digital Organizational Culture and Employee Wellbeing

Shyamasree Chakraborty
Research Scholar, Manav Rachna University
Dr Pragati Chauhan
Associate Professor HOD-Department of Management & Commerce,
Faculty of Management & Commerce, Manav Rachna University

Abstract

Worldwide, digital disruption is no secret nowadays. This is happening faster than we can think of. Every organization knows that technological disruption is continuously happening. Technology gave rise to and continues to fuel the digital disruption phenomenon. Still, the technological challenges companies are facing are significant in number. Many organizations are struggling, more specifically, about how to manage disruption, adjust to disruption, and prosper in a world within a finite time. However, the human mind is programmed to resist any type of change at first. Digital disruptions can create unknown fear among employees and employers alike. The fear of adapting to a new type of technology or of a new process can be very powerful. Technology will continue to interrupt all types of work, even if we do not know exactly how and when that interruption will happen. These can impact the mental health and wellbeing of an individual, and in turn affect the performance and output of an organization. Hence, organizations need to mitigate this fear among employees to thrive in these disruptions. Technology by itself can do little unless the people who work with it feel part of something special and valued. Change and innovation will always be met with fear or suspicion simply because it is different, and confidence and capability take a long time to establish themselves in our mind and body. The onus lies, both with employees and employers, to mitigate any unknown fear arising from technology disruptions and invest time and resources to thrive in this constant technologically disruptive world.

Keywords: *Digital Organization Culture and Employee Wellbeing, Impact of Digitization on Employee Wellbeing (Advantages & Challenges), The fear of Unknown: Impact on Mental Health & Wellbeing, Readiness approach to adapt organizational digital culture, Key takeaways*

Digital Organization Culture and Employee Wellbeing

Technology Innovation and digitization have taken a lightning speed. It has developed a behaviour of changing faster than most of us can handle. Post Covid 19, across all generations felt the desperate need of digital technologies both in personal and professional life. The need for innovation is driven by one of the basic natures of humankind, inquisitive mind, and its relentless attempt to react better in adverse situations. That's why the invention of the wheel around 3500 BC is considered as the biggest invention by humans and known as the hallmark of innovation.

The rapid advancement of digital technologies and the speed it has taken in the last two decades is making us shattered with both positive and negative effects. Percentages of positives are higher as it transformed our societies with better connectivity, commutes, simpler way for trade, superior financial ecosystems, globalized economy, productivity; in one word it united the world with "One-Touch".

Despite having a larger benefit, advancement in digital technology also led to some significant consequences on people's physical and mental wellbeing. With the faster progress, often people are getting confused to acknowledge and integrate constructively into our socioeconomic culture and life, which requires better digital readiness.

At the workplace as well, the story is quite similar. There is a constant force of technology disruption and adoption. Some may enjoy this continual change and new ways of working; however, the majority struggles the most in adaptation.

Growth and harnessing disruption are on top of each leader's agenda. As new competitors and new technologies shorten business cycles, generating new customer demand and new ways of reaching customers, there are no alternatives of digital resolution to meet the need. Nevertheless, employees also fear how this same technology will affect their livelihood, how the new tools & resources might reveal their shortcomings, how new tech-savvy competitors might poach the market space or possibly how the technology might eliminate the need for a specific role or job entirely. In these circumstances, digital transformation can be seen as something that stands between a person and their future and that, indeed, will be seen as a threat. Definitely this constant pressure of being future ready with skills and tools, compete with the world & deliver performance, add enormous stress for employees and have a major impact on wellbeing (mental and physical health).

In this chapter, we will discuss about:

- Impact of Digitization on Employee Wellbeing (Advantages & Challenges)
- Fear of unknown & its impact on wellbeing
- What things employees and employers both can do for better adoption and inculcate it in the organization culture.

It observed that the goal of digital transformation is promoted as a transformation of an organization's technological shift. However, in true sense, transformation must occur within each and every individual at an organization with whom the technology comes into contact.

Impact of Digitization on Employee Wellbeing

Technology transition and digitization is the necessity for every organization to survive in this competitive space. "Adapt, Advance or Die", the current landscape of the world, applies to all businesses, but it's easier said than done. Story of Nokia's rise and fall is the biggest example of why continuous advancement, adaptation and embrace changes are key to survival. Each one of us has witnessed the importance and advantages of digitization at the workplace. Following are a few advantages of digital transformation of workplace:

i. Ease of information sharing and connectivity:

Almost every human activity has gone through a change transition as a result of advancement of our communication and information sharing technology, provided better connectedness across the globe. At workplace, this revolution is associated with continuous innovation and investment towards digitization. Online access to unlimited information & AI based tools/technologies enabled employees to acquire and process information faster. Through the advanced technologies, an organization's culture is becoming more fluid, flexible and increasingly interconnected. This has resulting into higher collaboration amongst employees and better conflict resolution across level & geography. When an organization is more connected and collaborated, it reflects a healthy culture.

Disruption leads to new opportunity and sustainability

Change offers opportunity in abundance if taken with the right spirit and have the appetite to learn new things. Digital organizational culture opens a plethora of possibilities to shift from traditional mundane job roles to the new 'boundaryless' career options (Rodrigues and Guest, 2010). Consistent investment in digital organization culture means the job roles and responsibilities are often redesigned to beat the competitive market, subsequently it brings superior expectations for employee's re-skills for different roles across their career. This leads towards individual growth and improves financial stability. Re-skilling and getting ready with the digital changes bring sustainability for individuals, resulting to greater cognitive response.

Digital Organization culture boosting diversity & inclusion (D&I)

Technology & Digital workplace culture plays a crucial role to embrace the diversity and have a better inclusive environment for employees to shine. This results in better organizational productivity, success, and employee wellbeing. Advanced technologies equipped the organization to implement D&I strategies effectively with larger impact. It provides the opportunity to attract talents from all over the world and can radically enhance organization culture for employees of all backgrounds, regardless of race, gender, religion etc. Technology advancement emphasises more on transparent and open processes, leads to less biases and a more trust-worthy environment within the workforce. Through the integration of new technology into the critical talent management processes like Talent acquisition, Talent development & training, Talent performance management and effective communication. This provides a superior understanding of unconscious biases around D&I and influences the organization culture to be more people centric.

Employee happiness, satisfaction and wellbeing are crucial for an organization to deliver successful business results and D&I has a significant role to play here. Employees who get satisfaction in doing their work form a great example to others who are less Happy. Employee Happiness and satisfaction are subjective concepts, however it is observed that a diverse and inclusive environment at work is an essential requisite for deriving performance. There is a correlation between inclusiveness and happiness on the job.

Most of the technology expansion has multiple advantages, however it can have unwanted disadvantages too. If not taken into care during transition, it may lead to complete failure of the activity. A few of them are:

- **Change resistance:** We are never too old to adapt new. However, our dominant psychological tendency always plays its role whenever there is a change. Our brain is prone to react negatively to change. Negative bias commands more attention and it spreads faster. Learning and reskilling are the core of digital disruption drive for an organization, but widespread negative perception around the change may become the biggest obstacle for adoption. Adoption refers to the difference between the level of digital change increase and the rate at which individuals are making these changes a part of day-to-day life. Digital organization culture transition is just not a top-down approach. This requires creating an environment where every individual is thinking and working differently to adopt the change. This requires mind-set change and acceptability across every level and geography.
- **The gap in awareness and action:** The rapid rise of digital technology and the demand it created in the market space, despite knowing its importance there is a gap in awareness and action. The gap is larger than anticipated. Being aware of digitization happening at organization and doing something to get ready to meet the future skill requirement are two different aspects. For every organization, there might be a well-developed strategy & plan of implementation at the CEO's desk, however the reality could be different till the time those strategies are not actioned at ground level. An organization is considered to be an effective organization, when organization culture is mould with the effective communication/responses to that disruption needs and involve both talk and action. Digital organization culture is a blend of technology and individual readiness. Readiness means knowledge and daily actions, both. Until the time individual in the organization is not emotionally and intellectually cultivating a culture that allows to digitally ready, simple talk would not suffice the behaviours of true digitally mature culture.
- **Competency trap:** In management dictionary, competency traps are described as beliefs that the factors associated with past success, would also lead to future success. Traditional organizations typically face substantial challenges when it

comes to technology and digital transformation. Delivering values through work to the customer and new services in the changing competitive landscape, past success might not serve the future requirement.

GE is an example of the same. The organization is still overcoming competency traps. During the 1990s, GE was renowned for its practice to a process known as Six Sigma. Six Sigma is a quality management procure for controlling the error rate in manufacturing processes to 0.00033%. This six-sigma quality measure was a key element associated with GE's success during that period. However, this process also has its limitations. To maintain the standards of six sigma, the process itself is quite strict and not favourable to the types of agile responses to environmental shifts that characterize the world of digital business. To address the need for faster change and agility, GE has developed a complementary approach known as *FastWorks*, which combines lean start-up principles created by business consultant *Eric Ries* with GE's size and resources. As GE's culture transformation leader Janice Semper notes, "It's been harder in some of our businesses where they don't have as much volatility or disruption yet. They don't necessarily feel the same need as those parts of our businesses that are being disrupted or are in extremely challenging, volatile, ambiguous environments where this is really the only way to go forward and work."

The Fear of Unknown: Impact on Mental Health & Wellbeing

From the perspective of evolution, humans have evolved as complex creatures. We are continuously evolving. This evolution has imparted substantially for development of our knowledge base, our thinking around our past and future. Similarly, we regularly have the irrational and instinctive fear around us related to unknown factors, which have been with us in the whole evolution process. We create the fear both inwardly and outwardly, and it is like a regular partner staying with us in our life. Most of the fears we regularly have do not create abject terror, or nightmares. However, this fear can also present as hesitancy, deferment, grief, annoyance, or rebuttal in our behaviour. Fear can be painful, disturbing, even absolutely devastating.

Fear can also emanate from a stressful situation, which in turn initiates a torrent of stress hormones in our body to produce physiological changes. A stressful incident can make our muscles tense, beads of sweat appear with increasing heart beat and quicker breathing. This is basically a survival mechanism which is also known as the "fight-or-flight" response. It enables humans to react quickly to life-threatening situations. The poison of fear transmits itself quickly across individuals, groups. When we witness or hear someone voicing fear or concern, we immediately fall back to our default setting. When it becomes a shared fear, it can transmit through people across demography. Then the fear can stop any progress or change happening in its tracks and sometimes even reverse the change.

The goal of digital transformation is promoted as a revolution of a company's technological workings. However, the true transformation must occur within each employee with whom the technology has to interact. Now, as mentioned earlier, change is not welcomed easily by the human brain, and as such there will always be resistance to the change. The resistance can create fear in the mind. The corresponding fear may displace a person from the apparent security of their work.

Employees may think about the implied threat of job loss or punishment that comes with them if they cannot adapt to the technology changes. People will simply be in a hurry to get their tasks done, and they may have little time or energy available to deploy skills such as critical thinking about how to familiarize with digital transformation.

Readiness approach to adapt organizational digital culture:

As employers aim to push growth through innovation in the face of significant technological or digital transformation, impact of change at employee and employer level will be inevitable. Instead of impending change as a project to be delivered, employers should focus on embedding long-term change-readiness into organizational culture and integrate with the value system - beliefs, decisions, behavioural expectations, and capabilities. Technology transformation and culture are typically considered as two different subjects and managed as separate projects. But in reality both are entangled with each other and require change management to deliver successful results.

When employees exhibit behaviours that will soon be obsolete due to digitization, or when a culture and supporting programs, initiatives and infrastructure are aligned with an obsolete business model, it becomes difficult for organizations to take necessary actions for successful digitization or technology transition. How technology transition and organizational culture relate to a future success or failure, depends on an organization's enduring change-readiness and appetite to evolve with new technology.

Digital readiness or to reimagine the future requires both organizational and individual readiness and upskilling.

Organization Readiness

Organizations struggle with knowledge of the ingredients to make a digital-first culture. Digital culture runs deeper than redesigning the workspace with digital gadgets and changing the name of the job roles. Organization's readiness depends on how an organization behaves, what it values, its unspoken but deeply embedded beliefs. Digital culture is often described as being in the "DNA" of the culture.

Organizations must have severed strategies and actions aligned to build momentum for the transition, however these strategies and actions to be embedded as part of the DNA of the culture, it requires effective leadership. An effective leadership is one of the most critical aspects associated with digital readiness for an organization. This involves leadership vision, decision making capacity, agility, and power to build a connection with employees.

Leadership vision: Leaders are for better tomorrow. Employees look at the organization's future through a leader's lenses. A clear and futuristic direction helps employees to connect with organization goals and purposes. It embalms and establishes the base of digital organization culture through open/transparent, long-term-oriented, proactive growth-focused and a clear vision. To become future ready and high-performing organization, strong and effective leaders always keep a check of:

- Employee understanding of the mission, vision, and purpose of the organization;
- How foster the highly skilled competence and nurture mastery;
- Employee belongingness: Nurture a sense of belongingness to at work;
- Employee autonomy, in both professional and personal lives;
- The articulation and communication of a clear vision and creating excitement about the future;
- As a leader, how are they explaining the rationale for change and having a structural communication reach every individual
- Being open and honest;

- How this change impacts employee wellbeing (mentally and physically).

Effective Decision making: One of the biggest and critical tasks from every leader is "fast and effective decision making". Due to the faster rate of change at which digital disruption is happening, leadership does not have the luxury to take poor or slow decisions. Delay in decision making or ineffective approach to decide the way forward, may just push the organization behind in the competitive curve. Faster and agile decisions are the key characteristics of future and change ready organizations. Leaders who are strong in decision making, consider:

- Positive and negative impact of the decision.
- Employee emotional and intellectual impact
- Consult and involve right people during decision making
- Fact and data driven decisions
- Value creation through the decisions taken

Agility: During any transition one of the major challenges that organizations face is keeping a continuous rhythm and adapting to all aspects of the changes with the same speed that the change is happening in the world. Hence, digital strategy and implementation is not necessarily a one-time long-term plan to deliver followed by a multiyear implementation timeframe. Rather, it should be a continuous process of identifying the overall objective and goal of digitization in the business. In the face of constant change, knowledge of specific skills and routine behaviours will soon be obsolete. The only sustainable capability is knowing how to learn and adapt. This requires having short-term actions that get the organization closer to the digitization goal and provide opportunities to rethink and learn from the outcome of the short-term initiatives. Agility empowers organizations to challenge the status quo with open minds, course correct and be with the momentum of the digital progress.

Build Connect with Employee: Employees are the core of any organization's technology or digital shift. The growing distance due to virtual reality and post covid hybrid culture, despite AI/ML technologies are taking centre stage, every successful organization and leadership redefining their strategy to build emotional connect with employees. They understand the intensity and seriousness of having a strong employee relationship to build a resilient organization. Leaders with strong employee connections build an environment around them where each employee feels safe and included. Employees are empowered to provide constructive feedback and participate in the change journey.

A positive employee experience within the workplace is a driving force to facilitate successful growth and having a podium to nurture healthy wellbeing. This will, in turn, build a trust and high performance led culture for the organization to thrive. Although we can vary many rules according to the culture and technology transition needs, leaders with absolute vision and right intent, sets them apart by (i) Giving everyone a voice and (ii) weaving core values into everything. Values that hold to be so fundamental that everyone holds & demonstrate them irrespective of whether they are rewarded.

Individual Readiness

Individual readiness for a change or to be future ready with the advancement of technology, is the readiness in the beliefs, attitudes, and intentions of employees that are comprehensively and consistently influenced by the transition process, context, and objectives involved in the change movement.

A regular person spends around 90,000 hours of their life at work. And, not before spending their first 20-22 years (that's 192,000-plus hours, right there) to prepare to get ready to join the workforce, garnering education, and other experiences. A major part of our lives, then, is spent at work or working to get work. It is only natural that the way we feel about our workplaces and its culture shapes so much of our outlook. How we feel at work has a direct impact on how adjusted or even, happy, we are with our lives or the overall wellbeing. Unfortunately, a lot of us don't feel good when our regular status quo or comfort at work is in threat due to change. This creates discomfort and disconnect at the workplace.

Human brain is wired to react more emotionally to find fault in others when there is discomfort or disconnect. Our natural change resistance tendency despite acknowledging the fact that change is necessary, comes between adoption of new ways of working and embracing it wholeheartedly. This leads to an inverse impact on oneself. This affects our mental and emotional health, and we get into a loop of continuous criticism and comparison. This further dent our individual progress, growth and become future ready with new skills and competencies, which in turn, make professional careers stagnant. This automatically starts affecting the wellbeing of individuals and pushes towards depression and mental health issues.

Elisabeth Kubler-Ross one of the most known social psychologists in her influencing work "On Death And Dying" (1969), she described five stages of grief. However, this model is not only applicable to the trauma of death, but it is observed that this also works in a similar pattern when changes occur in an individual's life. In fact, this is considered as one of the defense mechanisms or coping mechanisms that an individual has to move through in order to manage change.

In the Kubler-Ross model, she describes the five stages as: denial, anger, bargaining, depression, and acceptance. [Figure 1]

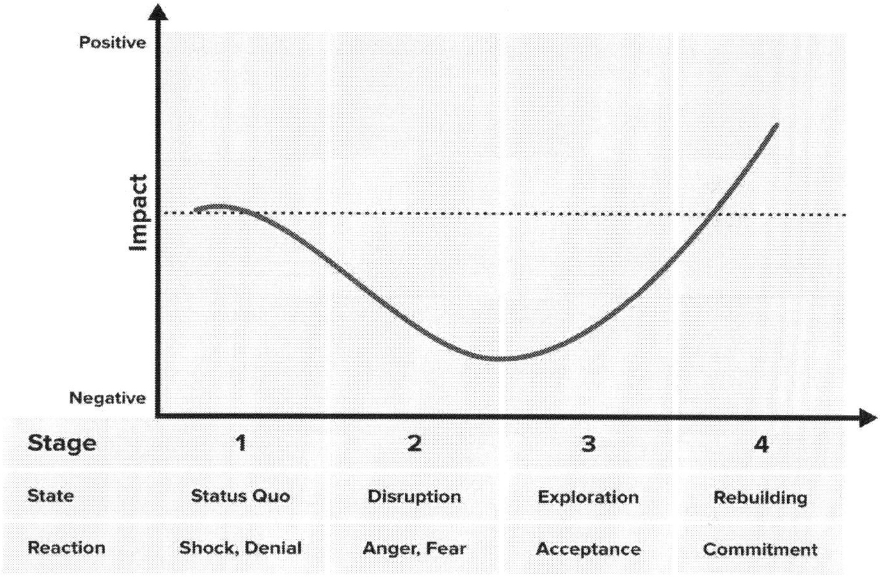

Figure 1: The Kubler-Ross Grief/Change Model

The question is how employees can own the responsibility to be future ready and seamlessly transition through each stage of change management. This requires behavioral and intellectual shifts.

- Will (Intent): Deals with the motivation and willingness of the individuals
- Skill (Ability): Deals with the ability and competence of the Individual
- Build (Awareness): Deals with the personality & traits of the Individual

Until we are not self-sufficient and have a desire to adopt the new normal, no one can make us ready to face any change at the workplace. Our 'will' or 'intent' help us to get more engaged and involved in what we want and how uniquely we want to achieve the same. True potential is realized when individuals are aware of their thoughts, feelings & behaviors, and approach changes with a sense of belongingness & clarity. When individuals operate with this mindset, they tend to give the best in every pursuit.

References

Anya Johnson, S. D. (2020). A review and agenda for examining how technology-driven changes at work will impact workplace mental health and employee well-being. *Australian Journal of Management, 45*(3), 402-424.

Author, N. (2020, July 6). *Understanding the stress response.* Retrieved July 2022, from health.harvard.edu: https://www.health.harvard.edu/staying-healthy/understanding-the-stress-response

Bolli, T. P. (2022). Decomposing the effects of digitalization on workers' job satisfaction. *International Review of Economics*(69), 263-300.

Gerald C. Kane, A. N. (2019). *The Technology Fallacy-How People Are the Real Key to Digital Transformation.* Cambridge, Massachusetts: The MIT Press.

Lopes, A. B. (2001). The Knowing-Doing Gap. *ADMINISTRATIVE SCIENCE QUARTERLY, 46*(3), 558-559.

Prentice, S. (2022). *The Future of Workplace Fear: How Human Reflex Stands in the Way of Digital Transformation.* Toronto, ON, Canada: Apress.

Ricardo A Rodrigues, D. G. (2010, April 30). Have careers become boundaryless? *Human Relations, 63*(8), 1157-1175.

Rozin, P. R. (2001). Negativity Bias, Negativity Dominance, and Contagion. *Personality and Social Psychology Review, 5*(4), 296–320.

Williams, M. (2022, February 3). *How technology can boost diversity and inclusion in the workplace.* Retrieved July 2022, from www.thehrdirector.com: https://www.thehrdirector.com/features/diversity-and-equality/how-technology-can-boost-diversity-and-inclusion-in-the-workplace/

❑❑❑

SECTION - 2
Legal Sector

5

Digital Banking and the Right to Privacy: An Insight into India's Legal Position

Chaitali Wadhwa
Assistant Professor, School of Law, Manav Rachna University

Abstract

The banking industry has seen considerable growth in digital banking tools. Data generated through banking is synonymous with a new business energy source. Digital banking has proven its worth during the COVID pandemic. Despite its many advantages, the growth of digital banking is not without its challenges and limitations. Among others, one significant concern is the security and privacy of information. Banking remains one of the most at-risk sectors when it comes to privacy violations and data protection. In India, there have been many cases that have discussed data protection and the right to privacy at great lengths. With advancements in technology, the laws have also evolved and expanded to include several segments within their ambit. The purpose of this paper is to address these concerns of privacy violation in the era of digital banking. The author sheds light on some legislations and draws reference to landmark judgements to trace the evolution and reformation of the legal stance on data protection and privacy.
Keywords: *data protection, right to privacy, digital banking*

Introduction

The internet acts as a facilitator in many aspects of life – be it interactions with other people, conducting businesses, assisting in electronic commerce, data management, and provision of services.

One such industry that has benefited greatly from the growth of technology and the internet is the banking industry. The banking industry has seen considerable growth in digital banking tools, such as sending money, shopping online, opening and managing accounts, and applying for loans. Digital banking has proven its worth during the COVID pandemic (Jayaraman, 2021; RBI, 2022). It allows businesses to effectively interact with customers, provide any-time services, and manage complex challenges (Omariba & Masese, 2013).

Despite its many advantages, the growth of digital banking is not without its challenges and limitations. Among others, one significant concern is the security and privacy of information. Data privacy and digital banking have become a hot topic in the current social and economic scenarios, especially with the rise of cybercrimes and data collection by business giants. Banks and financial institutions collect sensitive information from customers, on the pretext of curating a better marketing experience. The breach of sensitive data may have dire consequences. Storing information on the cloud has made access easy but brings along with it concerns of data privacy (Swinnen, 2020).

The paper aims to provide a legal perspective on the right to privacy in the context of digital banking. It begins by tracing the growth of digital banking. Part III of the paper addresses concerns about digital banking, putting special emphasis on privacy violations. The next segment provides insights into the legal framework in India. Part V is the conclusion.

Growth of Digital Banking

India's fintech market, estimated to be $31 billion, is ranked at number 3 in the global scenario (Conventus Law, 2022). The Reserve Bank of India (RBI) has played a significant role in transforming and developing India into one of the most modern payment systems in the world (RBI, 2022). Some of the services offered under the umbrella of fintech include digital banking, mobile banking, unified payment interface (UPI), blockchain, chatbots, cloud banking, and biometrics. UPI is one of the fastest ways to provide real-time banking transactions at any location and time. In India, UPI is developed by the National Payments Corporation of India, and it is regulated by the Reserve Bank of India.

Blockchain technology combines several technologies to store large amounts of databases and streamline complex working processes. Blockchains operate in both public and private forums, allowing customers more inclusive and more secure transactions (Portilla et al., 2021). The United Nations and the International Monetary Fund (IMF), as well as several countries including the United States, United Kingdom, and Japan, have been exploring the application of blockchains in the financial sector (Guo & Liang, 2016). Niti Aayog has created India's largest blockchain network named IndiaChain. In addition to the above, the Government of India and the RBI are encouraging the growth of fintech startups.

Three significant waves transformed digital banking in India (Ahmed & V. V., 2020). Wave one was the conversion to electronic operations through the use of ATMs, call centres, and telephone banking. The second wave focussed on the growth of SMAC technologies, i.e. Social, Mobile, Analytics, and Cloud technologies, that enable financial institutions to promote personalised banking services. In the third wave, there has been a rapid increase of new technologies such as Artificial Intelligence (AI), Blockchain, and Robotic Process Automation. Together, they provide enhanced customer experiences.

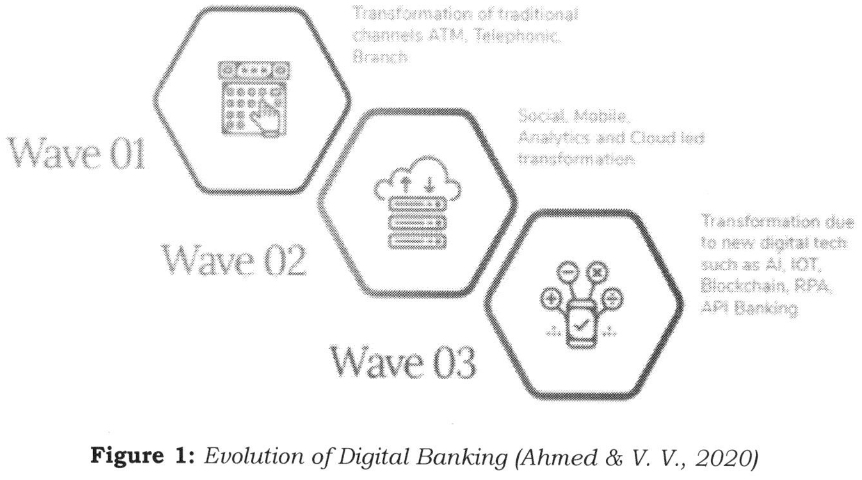

Figure 1: *Evolution of Digital Banking (Ahmed & V. V., 2020)*

Initially limited to banking channels, digital technologies have now encompassed the entire banking spectrum, ranging from back-end operations to customer interface (Ahmed & V. V., 2020). Shifting from traditional forms of banking to digital banking has changed the lifestyle of Indian citizens. According to a recent study, the number of consumers preferring mobile banking has increased from 47% in the pre-pandemic era to 57% now (Jayaraman, 2021).

Concerns about Digital Banking

Continuously evolving technology has paved the way for transformations in digital banking but it does not come without its challenges and concerns. For uneducated individuals and senior citizens, it is difficult to rely solely on the digitization of all processes. The apprehension to go cashless stems from a well-founded fear of making a minor error that may result in major financial loss. Though digital banking provides a wide array of services and products at low costs, many individuals still feel the comfort of traditional banking (Liao & Cheung, 2003). Digital banking has the potential of resulting in cases of cyber frauds and cybercrimes. Cyber security continues to remain one of the major issues in the field of digital banking. Looking at the various concerns that arise with digital banking, it is safe to say that the greatest challenge is not the application of new technologies, but rather the enhancement of consumers' acceptance (Liao & Cheung, 2003).

Though the advancement of information technology has assisted in streamlining the collection and analysis of data, it has also created vulnerabilities. In India, specifically, the vulnerability lies in the lack of clarity regarding the classification of information (Basu, 2010). Banking remains one of the most at-risk sectors when it comes to privacy violations and data protection, as it involves the exchange, recording, and retention of sensitive, highly personal data (Hickok, 2010). Examples of privacy violations in banking include, but are not limited to, sharing an individual's personal information with third parties, accessing credit history, and loss of data due to improper security.

Right to Privacy

The idea of privacy is intricately woven with the foundations of *"liberty, justice, human dignity, individuality, and family life"* (Basu, 2010). The idea itself has been long standing in society, but its codification as a right is a mark of recent trends. As societies transform, the idea of privacy as a right is also re-conceptualised and re-defined. While defining the right to privacy, researchers tend to ignore specific societal, political, economic, and socio-historical contexts, and employ biases in favour of the West (Basu, 2010).

Ruth Gavison cites three components that form a complex combination to constitute the right to privacy – *"secrecy, anonymity, and solitude"* (Gavison, 1980). Though the elements are independent of each other, they are related when the context of privacy is observed. While addressing the right to privacy in a social framework, it is imperative to associate with the definition of an individual's *"reasonable expectations"* that give space for changes in technology and culture (Basu, 2010). In India, therefore, the right to privacy entails personal space and subjects (Kumaraguru et al., 2005). In the 1977 judgement of *Ganeshi Lal v Rasool Fatima* (AIR 1977 All. 118), the court recognised the protective approach of women towards privacy in their homes. Similarly, the Allahabad High Court in the case of *Basai v Hasan Raza Khan* (AIR 1963 All. 340) extended the customary right of privacy in cases of *"purdahnashin"* women.

Subhajit Basu distinguishes the two types of privacy – *"informational privacy and decisional privacy"* (Basu, 2010). The former deals with the use, transfer, and processing of personal data. On the other hand, decisional privacy comes into play when there is interference while making certain fundamental decisions.

Legal Position in India

India's legislative history is indicative of the country's general unwillingness to adopt data privacy laws (Basu, 2010). Privacy laws in India were assumed to be *"notoriously weak"* because of the absence of comprehensive legislation (Basu, 2010). The existing position is quite contrary to this assumption, which rests on a paradigm ignoring the element of "trust", prevalent in the very culture of Indian society.

The growth and development of the laws are an amalgamation of constitutional machinery, customary rights, and common law influences. Society's attitude also shapes the formation of the laws and judgements passed by the Courts. The judgement passed by the Delhi High Court in the case of Naz Foundation (*Naz Foundation v Government of NCT of Delhi & Ors, 160 (2009) D.L.T. 277)* is indicative of the right to privacy imbibed in the term "dignity" as laid down by the Constitution of India. The first time the right to privacy was recognised in India by an eight-judge bench of the apex court in the judgement of *M P Sharma v. Satish Chandra, District Magistrate, Delhi* ((1954) SCR 1077), and then again by a six-judge bench in *Kharak Singh v State of Uttar Pradesh* (AIR 1963 SC 1295).

In the year 1997, privacy became a matter of discussion when a civil rights NGO – PUCL – filed a petition in the Supreme Court, based on an allegation by parliamentarian Chandra Shekhar Singh that the then-government led by V P Singh was illegally tapping the telephones of politicians. The allegation snowballed into a national scandal, and a report by the Central Bureau of Investigation exposed that phone tapping was regularly carried out (Ramachandran, 2014). The court in its judgement *People's Union for Civil Liberties v Union of India* ((1997) 1 SCC 301), affirmed that phone tapping was a violation of the right to privacy. The court also laid down guidelines to act as a safeguard against the arbitrary exercise of the executive's surveillance powers.

The issues around the right to privacy arose once again to challenge the constitutional validity of the Aadhaar Scheme by the Unique Identification Authority of India (UIDAI). The above-mentioned cases focus more on the physical concept of the right to privacy. The Aadhar judgement takes it a notch further by addressing concerns arising on a digital platform. The petitioner, Justice K S Puttuswamy, questioned the norms for the compilation of demographic biometric data by the Indian Government. Arguments by both parties revolved around the ambiguity of the right to privacy, and its crystallization under common law. The position of the court was affirmed in the historic landmark judgement of *Justice K S Puttuswamy (Retd) & Anr. v Union of India & Ors.* (AIR 2017 SC 4161), where a nine-judge bench recognised the fundamental right of privacy as a facet of Article 21, and Part III of the Indian Constitution. The Supreme Court overruled the previous judgements of M P Singh and Kharak Singh where they held that privacy was not a fundamental right. Other judgements by smaller benches too have affirmed the existence of the right to privacy in the Indian legal framework.

Due to the rising concerns, the fintech industry has shifted from a *"business centric approach"* to *"compliance centric approach"* to privacy and security (Conventus Law, 2022). A *"user centric approach"* requires that fintech companies should integrate privacy and security as a regular check to maintain users' confidentiality and integrity. The primary legislation protecting data in India is the Information Technology Act, 2000 read with the Information Technology (Amendment) Act, 2008, and the Information Technology Rules, 2011. The amended laws through Sections 43A and 72A address the lacunae of data protection laws. Companies must now maintain reasonable security practices and procedures with respect to sensitive personal data and information. To provide a stricter and refined set of guidelines, the Lok Sabha introduced the Personal Data Protection Bill in 2019. The onset of this data privacy regime was influenced by the enforcement of the General Data Protection Regulations in the European Union. As of 2022, the Bill stands withdrawn, to pave the way for a *"comprehensive framework of contemporary digital privacy laws"* (Business Standard & PTI, 2022).

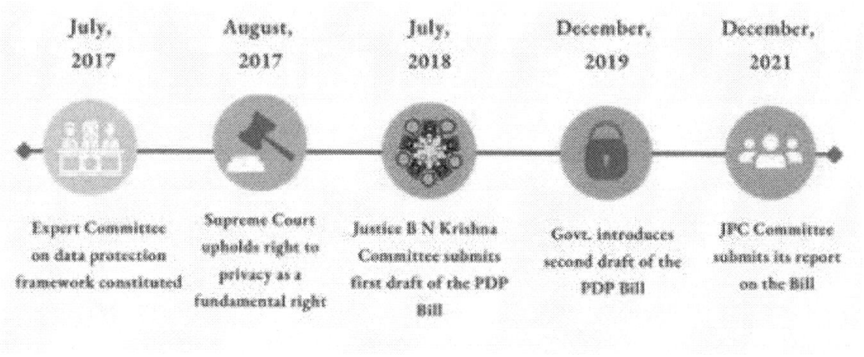

Figure 2: *Development of data privacy framework in India (Conventus Law, 2022)*

Apart from these, the issues of confidentiality in bank transactions fall within the ambit of the Recovery of Debts due to Banks and Financial Institutions Act, 1993. Similarly, the Telecom Regulatory Authority of India protects privacy in telecommunications. In addition to being guided by a legislative framework, many service providers have also voluntarily engaged in self-regulation to reduce the misuse of personal data (Basu, 2010). The Data Security Council of India has been entrusted with the task to generate awareness of privacy and security issues.

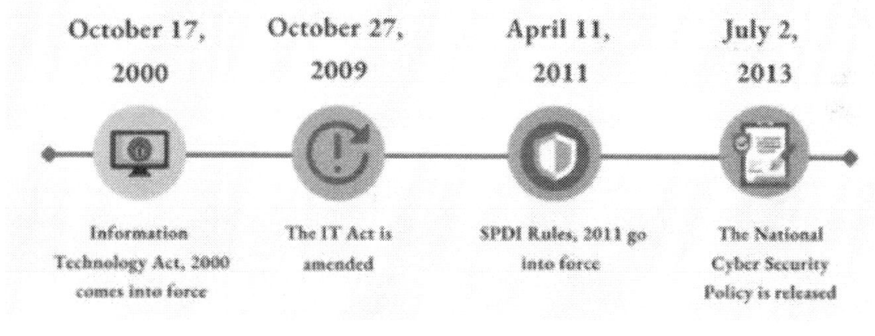

Figure 3: *Development of Indian cybersecurity framework (Conventus Law, 2022)*

Conclusion

The purpose of this paper was to address the concerns arising with digital banking and provide a legal insight into the framework in India that protects the citizens' data. The right to privacy has emerged from the customs and found its way into the constitutional framework that forms the bedrock of the democratic structure. Emerging technology in transforming society leads us to beg the question of whether a model legislative framework can resolve the concerns of privacy that arise with new introductions to existing digital structures.

The right to privacy has come to be recognised as one of the fundamental principles governing citizens, and the role of the Court has addressed the public's problems. Despite its criticism and limitations, the current legal position in India has done a decent job of expanding the definition to encompass data protection on a digital platform. Perhaps only time will tell whether new grievances will give rise to new solutions and transform India's position on a global level.

References

Ahmed, S., & V. V., S. (2020). Digital Banking in India-Recent Trends, Opportunities and Challenges. *Global Journal for Research Analysis*, *9*(1), 1–4. https://doi.org/10.36106/gjra/5605392

Basu, S. (2010). POLICY-MAKING, TECHNOLOGY AND PRIVACY IN INDIA. *THE INDIAN JOURNAL OF LAW AND TECHNOLOGY*, *6*, 65–88. http://www.ico.gov.uk/upload/documents/pressreleases/2008/rsa_speech_oct08_final.pdf.

Business Standard, & PTI. (2022, August 3). *Govt withdraws Data Protection Bill, 2021, will present new legislation.* Business Standard News. https://www.business-standard.com/article/economy-policy/centre-withdraws-personal-data-protection-bill-2019-to-present-new-bill-122080301226_1.html

Conventus Law. (2022, June 14). *India - Interaction Of Fintech With Privacy And Security _ A Tale Of Changing Priorities.*

Gavison, R. (1980). Privacy and the Limits of Law. *Yale Law Journal*, *89*(3), 421–471. https://doi.org/10.2307/795891

Guo, Y., & Liang, C. (2016). Blockchain application and outlook in the banking industry. *Financial Innovation*, *2*(1). https://doi.org/10.1186/s40854-016-0034-9

Hickok, E. (2010, November 22). *Privacy and Banking: Do Indian Banking Standards Provide Enough Privacy Protection?* Center for Internet and Society. http://www.cis-india.org/advocacy/igov/privacy-india/privacy-banking?searchterm=Bankin

Jayaraman, K. (2021). *Why should banks focus on data privacy while collecting customer data_ - Aspire Systems.* Aspire Systems. https://blog.aspiresys.com/banking-and-finance/digital-banking/banks-focus-on-data-privacy-while-collecting-customer-data/

Kumaraguru, P., Cranor, L. F., & Newton, E. (2005). Privacy Perceptions in India and the United States: An Interview Study. *33rd Research Conference on Communication, Information and Internet Policy (TPRC).*

Liao, Z., & Cheung, M. T. (2003). Challenges to internet e-banking. *Communications of the ACM, 46*(12), 248–250. https://doi.org/10.1145/953460.953507

Omariba, Z. B., & Masese, N. B. (2013). Security And Privacy Of Electronic Banking. *Kidney International Supplements, 3*(3), 262.

Portilla, D., Kappos, D., Ngo, M. Van, Rosenthal-Larrea, S., Buretta, J., & Fargo, C. (2021). Blockchain in the Banking Sector: A Review of the Landscape and Opportunities. *Harvard Law School Forum on Corporate Governance, 1*, 1–6.

Ramachandran, C. (2014). PUCL V. UNION OF INDIA REVISITED: WHY INDIA'S SURVEILLANCE LAW MUST BE REDESIGNED FOR THE DIGITAL AGE. *NUJS Law Review, 7*(2), 105–119. http://indiatoday.

RBI. (2022). *Payments Vision 2025.*

Swinnen, E. (2020). *The Ultimate Data Privacy Guide for Banks and Financial Institutions.* NG Data. https://www.ngdata.com/data-privacy-guide-for-banks-and-financial-institutions/

CASES CITED

Basai v Hasan Raza Khan, AIR 1963 All. 340

Ganeshi Lal v Rasool Fatima, AIR 1977 All. 118

Justice K S Puttuswamy (Retd) & Anr. v Union of India & Ors., AIR 2017 SC 4161

Kharak Singh v State of Uttar Pradesh, AIR 1963 SC 1295

M P Sharma v. Satish Chandra, District Magistrate, Delhi, (1954) SCR 1077

Naz Foundation v Government of NCT of Delhi & Ors, 160 (2009) D.L.T. 277

People's Union for Civil Liberties v Union of India, (1997) 1 SCC 301

6

Digital Courts of Digital India

Mudit Jain
Research Scholar, School of Management & Commerce, Manav Rachna University
Vishnupriya Pandey
Research Scholar, School of Management & Commerce, Manav Rachna University

Abstract

The Era of Information and Communication Technology (ICT) has transformed the lifestyle of human beings. The presence of ICT can be seen in almost all spheres of life from personal to professional. The present chapter focuses on the use of ICT in the Indian Judicial System. The Indian Judicial system has been suffering for a long time from the stigma of low disposal rate and pendency of cases. People were afraid to approach courts because of their inefficient, costly, obscure and lethargic way of disposing of cases. However, the introduction of ICT in the Indian Judicial System with the E-Courts Project (2005) opened the door for ICT in the Indian Judicial System thereby, slowly removing the drawbacks of the Indian Judicial System.. Furthermore, the Covid-19 pandemic witnessed a tremendous surge in the E-courts litigation in India. There has been a joint effort of the Supreme Court of India and the Government of India to introduce ICT in the judicial system of the nation, with the Supreme Court giving direction for introducing ICT in Court procedures and the Government of India investing money in equipping the courts with state of the art digital machinery.

Keywords: *Information and Communication Technology (ICT), E- Court Project, E-Court, Judicial System*

Introduction

Statements like "*Justice Delayed is Justice Denied*" given by British Prime Minister William E. Gladstone, looks suitable for the

Indian Judicial System. Courts have always been considered as temples of justice, wherein an individual seeks to get justice for the wrong caused to him. Judges are viewed as protectors of their rights ensuring justice is being delivered in the society.

Judiciary in any State has the primary function to deliver and ensure that justice is being done to the individuals, preventing infringement of the right of any individual, punishing the culprit for the wrong committed by him and providing relief to the individuals. Judiciary also acts as a pioneer for bringing social change through its judgments and throwing away evil practices within the society. It keeps a check on the powers exercised by other organs of the state i.e. executive and legislative, thereby preventing the government from becoming despotic. The Indian Judicial System suffers from a lethargic way of functioning.

There are several District Courts functioning under the supervision of 25 High Courts and one Supreme Court in India. Apart from these pillars of adjudication there are several tribunals and lok adalats for solving cases. There are about 5 crore cases pending out of which about 10% are pending for more than 10 years. Apart from lack of availability of courts, the Indian Judicial System also suffers from inadequate no. of Judges. Giving a reply in the Rajya Sabha Hin`ble Law Minister Mr.Kiren Rijiju mentioned that Judge Population Ratio in India stood at 2103 judges per million people in 2020. He also mentioned that the sanctioned strengths of the Supreme Court, High Courts and district courts have increased from 31 in 2014 to 34 in 2020, 906 in 2014 to 1079 in 2020 and 19518 in 2014 to 24225 in 2020 respectively (Economic Times, 2021). This was also done with the view to increase the Judge Population ratio and reduce the pendency of cases in the nation.

It is pertinent to note that being a Judge allows an individual to earn respect in society; however, holding the position of Judge and carrying out judicial functions is not a child's play. Judges have to ensure that justice is delivered to the individual and parties are allowed to be given fair opportunities to present their cases. A wrong judgment will not only bring the judiciary to ridicule but will affect its image in society. Similarly, an ineffective, inefficient and lethargic judicial system adds a bad name to the judiciary. Individuals in India often try to exclude going to courts for seeking justice, being afraid of a delayed, costlier judicial process. In ***Husssianra Khatoon v. State of Bihar*** 1979 AIR 1369, the Court emphasized the Right to a quick trial of under-trial prisoners' right to life under Art. 21. Thereby, attacking the sloppy judicial administration system which prevented under-trial prisoners from enjoying their lives for years.

Need for E-Courts

The Judiciary plays a vital role in imparting justice in society. It is that pillar of administration which ensures that the culprit is being penalized for the wrong done and relief is provided to the victim. Furthermore, faith in the administration and governance in a state can be ensured only in the presence of a strong and effective judicial system.

However, there are certain challenges which are required to be taken into account which are not only affecting the efficiency of the judiciary but are also making it ineffective in imparting justice timely. Preparing and maintaining decades-long records of the case always remained a tedious task for the court management. In certain instances there has been loss or manipulation of records, leading to the acquittal of criminals and delayed decision-making. The Hon'ble Supreme Court in ***State of Uttar Pradesh v. Abhay Raj Singh*** held that *"if court records go missing and re-construction is not possible, the courts are bound to set aside the conviction."* Prior to the use of ICT in the judicial system, there was lack of transparency in the court proceedings and procedures often leading to judicial corruption and pronouncement of wrong judgements.

Digital filing of the case has allowed the lawyer/litigant to check the status of the filing, applications, order so passed, and date of next hearing of the case by the courts at the click of the button which prior to the use of ICT remained to be a tedious, costly and time-consuming process and thereby making difficulty in having access to justice. While making appeals, a huge amount of time was wasted in summoning the cases from lower courts leading to delay. Furthermore, the need for use of ICT in the Indian Judiciary, also becomes necessary as Cases are adjourned merely because affidavits filed several years ago were not restored or traceable. In order to overcome the aforesaid challenges faced by the Indian Judicial System and to make it effective, ICT was introduced in the Indian Judicial System thereby, leading towards the creation of E-Courts in India.

E-Court Project

The establishment of the E-courts committee can be visualized as a foundation stone towards the introduction of ICT in the Indian

Judicial System. The committee was established by the efforts of the Hon'ble Supreme Court of India in 2004 with the view of assisting computerization of the judicial system in the nation.

The E-committee is the governing body responsible for supervising the E-courts Project, which is envisioned in the "National Policy and Action Plan for Implementing Information and Communication Technology (ICT) in the Indian Judiciary-2005." E-courts is a pan-India project supervised and funded by the Ministry of Law and Justice, Government of India. Its goal is to transform the country's judiciary via court ICT enablement (Drishti IAS, 2022).

The project has been introduced by keeping following objectives in mind:-

- Ensuring Development, installation and implementation of efficient justice delivery systems in courts.
- Making processes automatic thereby ensuring easier accessibility of information to its stakeholders.
- Improving judicial productivity, both qualitatively and quantitatively, by making the justice delivery mechanism quite convenient, cost-effective, highly reliable, and transparent.

National Judicial Data Grid (NJDG)

Being a part of E-court Project, NJDG acts as a monitoring tool which is being used for identifying, managing and reducing the pendency of cases. It involves provision of real time data for the pending cases (Verma, 2018) The National Judicial Data Grid (NJDG) is a monitoring tool that can be used to identify, manage and reduce the pendency of cases. It provides real-time data for the number of pendencies. It also classifies pending cases on the basis of women and children (Verma, 2018).

Benefits and Challenges of the Present Digital Court System in India

The use of ICT in the Indian Judicial System has bestowed several advantages not only to the Judicial system of India but also to the litigants, pleaders etc. as well as there are few challenges which need to be catered.

Benefits

Talking about the benefits which the use of ICT can bestow upon the Indian Judicial System are as follows:-

- **Transparency in Judicial System:-** Indian Judiciary has traditionally been suffering from issues of transparency and accountability. Checking the status of the case remained to be a tedious task. Further, litigants and parties were never sure about the contents of the order or judgements so passed in the court proceedings until they got a certified copy of the same. Getting a certified copy from the court is a time-consuming process and it reduces the time for challenging such order or judgment.

Further, there have been instances of Corruption in the Judiciary. Judges took money from the party and moulded decisions in their favour. All such instances question accountability, transparency, fairness and integrity of the judges along with the judicial system. These instances reduce the faith of the public in the judicial system of India.

The use of ICT in the Indian Judiciary under the E-courts project has brought transparency to the judicial system as one can easily check the status of his case by visiting the court's website and download the judgements passed in the case rather than approaching the lawyer for doing the same. Further, ICT has introduced Virtual hearings whereby, parties who are unable to attend court proceedings can attend the same virtually. This ensures them that their case is being defended by the counsel engaged by them and adds accountability to the judicial system. Further, the facility of downloading orders and judgments via the courts website, enables litigants and parties to prepare for their next step before getting a certified copy of the same.

E-courts app etc. available on the google play store has made accessibility to the status of the case along with order/judgements at the click of the button and the same can be accessed via mobile phone. Further, the use of such applications is not very complex in nature for the general public.

- **Accessibility:** The notion of rule of law is completely pointless unless the people get access to justice. Access to justice provides bedrock for any democratic system. It simply means ensuring justice is provided easily in appropriate time through legal processes (Dubey,(n.d.)). The Hon'ble Supreme Court of India is often regarded as custodian of Fundamental Rights and generally approached in rare cases. However, approaching the Hon'ble Supreme court for the poor and downtrodden class of the society still remains a big challenge, as it sounds highly illogical and costlier for such an underprivileged class to travel all the way to Delhi just to file a case against infringement of their right. Preamble also strives to achieve justice social, economic and political to all the citizens which cannot be ensured without having adequate access to justice.

Use of ICT in the judicial system will act as a step to counter the aforesaid challenges ensuring accessibility to justice. Any person including litigant or party to the case suffering from any illness or any physical deformity or coming from an underprivileged part of the society will not be able to attend the court proceeding virtually. Talking about issues such as accessibility to justice for women, it is pertinent to note that women have always remained to be a victim of the lethargic and ineffective Indian Judicial System. They remained unaware about their rights, court procedures for exercising their rights including the filing procedures which causes them to suffer from atrocities. The use of ICT in the judicial system provides women with information about their rights and the procedures for enforcement of their rights. Further, use of video conferencing during court proceedings provides a sense of security to the women as it eliminates any chances of harassment while coming to the court for attending court proceedings. It eliminates male domination in escorting female prisoners to the court thereby, also eliminating chances of harassment by male police officials (Sandhya,(n.d.)).

- **Easy storage/preservation of court records:-** Indian judiciary has been overburdened with cases. There are a huge number of cases pending in the Indian Courts, some of them are pending for more than 10 years. Such huge compilation of pending cases increases difficulty in management and of cases and along with the bulky records attached with them. The E-court Project, includes Case Information System (CIS) which helps in solving the issue of record keeping and case management. It reduces the chances of misplacement of information as the same being carefully recorded and stored. The system allows the litigant to know status of the case, orders passed by various courts, date of hearing of the case etc. online, accessible to any part of the world. At the time of filing of the case the litigant is given a CNR along with acknowledgement. The CNR allows litigants to check the status of the case whether it has been accepted or rejected along with the details of the order through SMS or online via court`s website. The CIS system has made Judiciary, litigant friendly and transparent (Calcutta High Court, 2022). Further, lowering down the burden of maintaining huge and bulky records as the same are available electronically.

- **Ensures security of witnesses and litigants:-** In many cases especially in high profile cases, it is being found that witnesses either refuse or are prevented from attending court proceedings to give their statements. They are often being threatened about the consequences which may be death, destruction of property etc. or are being harassed by the other party. E-court mechanisms may help in dealing with such cases, allowing witnesses to give their statement without physically attending the Court. The E-court mechanism provides online links for joining the Court proceedings via apps like Ciscowebex etc. wherein witnesses can join the court proceedings from anywhere and give their statements without any fear. Thereby, ensuring their safety and security.
- **E-Courts system ensures environmental justice:** The E-court mechanism not only saves time and money in travelling but also by doing the same it helps in ensuring environmental protection. Since the litigants need not have to travel for attending the court proceedings it reduces the pollution being caused by the various vehicles like cars, bikes etc. for attending the court proceedings. Further, the use of ICT is reducing air pollution. E–filings and digital way of maintaining records reduces the dependency on paper in court proceedings. Litigants need not have to submit their pleadings in hardcopy as they can do the same by E-filing. It not only saves the usage of paper in court proceedings but it thereby also ensures environmental sustainability. Thereby, making the judicial system of the nation environmentally friendly wherein, courts play an active role as custodians of the environment.
- **Speedier Court Proceedings:** Virtual Courts enable the Judges to hear personally each case which might not be possible for them in case of physical proceedings. It allows delivery of judgements faster with few resources. Thereby, not only reduces the time for resolution of disputes but also reduces the issue of pending cases faced by the Indian Judiciary and ensures timely delivery of justice.

Challenges

It is correct to say that virtual courts bestow various benefits upon the judicial system of any country, like India where the judiciary suffers from the issue of huge pendency of cases with a low disposal rate. With the introduction of the E-court Project, Hon'ble

Supreme Court has taken steps in making the Indian Judiciary tech-friendly. However, there is still a long way to make all Indian courts truly technological friendly. Some of the challenges in making it come true are as follows:-

- **Complicated Process:** The E-court system no doubt has its advantages like its efficient way of conducting the business of the court. However, it still suffers from the issue of cumbersome and complicated processes. The process of E-filing is quite a complicated procedure i.e. the documents including pleadings and evidence are required to be in particular formatting fulfilling all the specifications so required for filing. Failing which the document will be rejected and the party has to rectify the defects before filing again. This makes E-filing a complicated and time-consuming process. Furthermore, there are certain evidences which cannot be filed electronically, especially in criminal matters involving murder done by using a weapon..
- **Lack of techno-legal expertise:** Litigants come from different economic backgrounds, ages and tech skills. Those who belong to poor economic backgrounds do not have computers or access to good network connections. They may not be having necessary skills for participating in court proceedings. Furthermore, older litigants face more difficulty in dealing with E-court procedures than young generation litigants, due to a lack of skill and expertise in participating in virtual court proceedings. It acts as a barrier to justice by preventing litigants from putting their client's case in the best way.
- **Chances of issues like Hacking:** Records about the court proceedings under the E-court system are generally being stored in certain applications and software which are generally not so secure. Thus, there always exists a risk that the confidential information stored in the record can be corrupted and erased via a virus attack. Virtual courts run the risk of lack of data privacy and thereby becoming subject to cyber crimes like hacking.
- **Lack of Funds:** Using ICT requires huge investment in electronic gadgets like computers and infrastructure supporting the use of ICT like wifi facilities. Furthermore, funds are also required to meet expenditures incurred on the maintenance of such computers and infrastructure. The E-court project`s success requires a lot of funds for equipping all the courts of the country with ICT.

- **Connectivity Issues:** Benefits of ICT can be ripped only if there exists good and efficient connectivity. Poor connection affects court proceedings leading to passover and adjournment of cases. At times the court`s website suffers from server errors and buffering while joining through the link provided for court proceedings etc. which affects the smooth functioning of the court procedures and causes unnecessary delays in hearing the cases. Litigants and parties residing in remote and backward areas of the nation may suffer from a lack of infrastructure facilities and connectivity issues. People without proper infrastructure and high-speed internet connection face difficulty in attending court proceedings. Thereby, making access to justice unaffordable and inaccessible.
- **Costly for Litigants:** Use of ICT requires stuff like a computer, laptop and good internet facility along with a sound environment for attending court proceedings remotely without any noise. However, purchasing such stuff is quite expensive for the parties. Often they require maintenance which further adds cost to the parties/litigants.
- **Difficulty in court proceedings:** It often confuses the judges in the adjudication of cases wherein, one party attends the court proceedings virtually while the other party attends physically. This often leads to bad decision-making and delays in court proceedings including delays in decision-making and becomes a time-consuming and cumbersome process. Furthermore, in virtual hearings, it often becomes difficult to check the authenticity of the evidence and witnesses so produced before the court virtually. The issue of authenticity makes court proceedings questionable.

Suggestions

As mentioned above there are certain limitations/challenges required to be catered for ripping the benefits of ICT in Indian Courts. Some of these challenges can be removed easily whereas some may require time for their solution. There are a few suggestions which can be used in dealing with such challenges and thereby, making the Indian Judiciary truly tech-friendly. Some of them are as follows:-

- Technology-based education systems shall be formulated. Allowing the people to have knowledge about using digital platforms, electronic devices like computers, laptops etc. enhance their accessibility to the virtual court proceedings and are not dependent upon the attorneys for knowing the information about the case.

- Universities and colleges providing law courses shall educate students pursuing law about various online legal databases like SCC online, using electronic devices like computers for making pleadings and filing documents in the court. Information must be disseminated about mobile applications designed for court proceedings and their functioning.
- Availability of high-speed internet services shall be ensured on a pan-India basis, especially in remote areas of the Nation.
- The Gurus/Teachers in the law colleges shall be well informed with the knowledge about ICT, online legal databases and mobile applications only then they can guide students about using the same in a much better way. If the teachers are not enlightened with ICT we cannot imagine students being illuminated with the knowledge of ICT.
- Gadgets like computers shall be made available to poor and downtrodden students on bailment. Wherein students are provided such gadgets at free of cost provided they have to return the same timely and without any damage.
- Seminars and training sessions regarding using ICT in court proceedings shall be conducted for court staff, judges and litigants. Thereby, making them well versed with the use of ICT in the Judicial System.
- Online courses on E -court proceedings shall be introduced and provided to the public at nominal charges. Thereby, making the general public acknowledge the various technicalities associated with E-court proceedings.
- Best practices followed in other parts of the world especially in developed states for conducting virtual court proceedings shall be adopted while using ICT in Indian Courts.
- Online court proceedings including the information provided on the court's website, e-filing procedures etc. shall be made multilingual. It allows the party to visit and use the court's website, mobile applications etc in their native language thereby, preventing the language barrier which made them dependent upon lawyers. It will further enhance accessibility of the parties to court proceedings and provide them with information about the status of the case status in an easier way.

- Virtual Court proceedings shall be made simple and should not be complicated especially for a layman so that access to justice for all shall become reality.

Conclusion

We live in the era of digitalization involving the use of ICT in every sphere of life. The use of ICT in the Indian Judicial System through the E-Court Project has been a great initiative taken up by the Hon'ble Supreme Court of India. Under said project, the Indian courts are being equipped with state of art computing machinery with the view to make court proceedings simpler, faster, transparent, and accessible with the primary objective to reduce the pendency and overburdening of courts with cases. The use of technology in Judicial proceedings would ensure the timely delivery of justice to the Indian people. However, there is still a long way to go, there are millions of cases pending in the courts which require speedier disposal without compromising the quality of judgment. Present E-court mechanism in India suffers from issues like bad connectivity, server errors, a complex procedure for e-filing cases etc. The implementation of E-court mechanism leaves room for improvement like making the E-court mechanism multilingual enabling the parties to understand court procedures in their native language, provision of high-speed internet facilities in remote areas shall be given priority, and awareness about the functioning of the E-court mechanism must be spread among public, litigants, court staff, judges and law students thereby, making E-court project successful in achieving its objective of ensuring timely delivery of justice in the society. The present E-court project is the first step toward transforming the Indian Judicial System digitally, which will produce fruitful results in the near future.

References

Economic Times. (2021, August 5). Judge-population ratio stood at 21.03 judges per million people in 2020: Law minister. The Economic Times. https://economictimes.indiatimes.com/news/india/judge-population-ratio-stood-at-21-03-judges-per-million-people-in-2020-law-minister/articleshow/85072861.cms.

Drishti IAS. (2022, May 30). Digitisation of Indian Judiciary.. https://www.drishtiias.com/daily-updates/daily-news-editorials/digitisation-of-indian-judiciary.

Supreme Court of India. (n.d.). E-Committee. Official Website of e-Committee. https://ecommitteesci.gov.in/

Verma, K. (2018, September 26). E-Courts Project: A Giant Leap by Indian Judiciary. JOAL, 6(1), 1-14. https://ojs.law.cornell.edu/index.php/joal/article/view/77/75.

Dubey, M. (n.d.). Problems of accessibility to justice and litigation in India. CyberLeninka..

https://cyberleninka.ru/article/n/problems-of-accessibility-to-justice-and-litigation-in-india.

Sandhya, B.(n.d.) Accessibility of Women to Criminal Justice System. JOURNAL OF DEMOCRATIC POLICING, 4, 43-99.

https://keralapolice.gov.in/storage/pages/custom/table/table-OtLgTw3o13qbX0ZXRhqVHsRRE.pdf#page=44.

7

Evolution and Revolution of Artificial Intelligence: A Critical Analysis

Himanshi Babbar
Assistant Professor of Law, Vivekananda Institution of Professional Studies, Guru Gobind Singh Indraprastha University

Shivam Gaur
Assistant Professor, School of Law, Manav Rachna University

Abstract

For science fiction enthusiasts Artificial Intelligence is not a new concept. Nowadays, AI has become more science and less fiction. The creativity and knowledge exhibited by AI systems is clearly visible to the world and concerns regarding IP protection have sprouted in the minds of people enforcing rights in relation to intellectual property. Thus, we should explore more deliberative ends of copyright and patent laws in connection with AI systems. Nowadays there are machines that create highly creative works that would definitely be eligible for copyright protection if they were created by humans. This calls for a re-examination of copyright standards for AI systems all over the world. Inclusive growth in the future will mean the inclusion of every form of scientific advancements. This means that even robots/machines/AI systems should be part of this growth. A world deployed on AI is a destiny chosen by humans and it will be preferable to be appropriately prepared for this age of science where it is not mere fiction. It is time for a new world renaissance. In the near future, there will be new evolutions by the AI systems and thus this paper tends to examine and revise the laws regarding IP protection.

Keywords: *Artificial Intelligence, AI systems, Intellectual Property*

Introduction

People have long looked for creative ways to complete various tasks with the least amount of effort. Technologies have been suggested and are consequently continually being created for all of this, one of which is Artificial Intelligence technology. Even though the idea of artificial intelligence was originally put forth in the 1950s, there is still a lot of research and development that needs to be done, especially in wealthy nations. The idea that humans could one day create robots that can think for them and act autonomously has been embedded in our culture and literature since the birth of civilization.

Understanding artificial intelligence (AI) is a challenge. Despite the fact that the idea is not new, it has recently acquired so much attention that it is now being discussed by academics, corporate executives, policymakers, and scientists. This is because AI is such a broad field and has an effect on almost every facet of contemporary life. Even the fourth industrial revolution is thought to be critically dependent on AI. Self-driving cars, voice assistants, and drones are all instances of how far technology has come in recent years. As its use has grown to be essential in the majority of technical applications, artificial intelligence (AI) has gained importance in the modern day. AI has improved our lives by influencing a number of industries, including transportation, health, and others. As a result, all nations now tend to automate the majority of tasks and reduce human involvement in order to maximise effectiveness and eliminate mistakes.

India is becoming a significant hub for artificial intelligence patent applications, and it ranks among the top nations in terms of publications in areas like computer innovation, foresight, and natural language interpretation. India has moved to the top of the list of nations filing AI patents, according to a recent WIPO report. The study found that India has been ranked ninth in the "first-to-file" race for AI patents since 2015, despite the fact that the country has experienced excessive growth in this field in previous years.

Evolution of Artificial Intelligence

Machines have been around for a while, but what's new today and what is truly propelling this shift is the development of machine learning techniques. Really, machines are no longer restricted to gathering explicit data in circumstances where a person might clearly explain a series of rather rational steps. Our daily lives have already been affected by machine learning, which unlocks smartphones with a quick glance or touch, suggests music we might like, and trains cars to drive themselves. AI mostly focuses on providing information to this machine learning. Artificial intelligence is technology that mimics human performance, usually by learning, making its own decisions, appearing to understand complex material, conversing with people naturally, improving human cognitive ability, or taking the place of people when performing mundane jobs.

Technology that is created by a human, improved by a machine, and potentially disruptive to almost every aspect of human life is referred to as artificial intelligence (AI). As of now, technology is capable of acting rationally, thinking like a person (strong AI), or even outperforming humans (i.e., super intelligence). Systems in computer science that exhibit intelligent behaviour and are focused on creating intelligent machines are referred to as AI. Recent news reports have cited artificially made artwork and drugs found by computer programmes. The world has significantly changed as a result of the 270 percent growth in the number of companies utilising AI technologies over the past four years. AI is already having an impact on how goods and services are produced and distributed. The healthcare industry has been greatly impacted by AI, which has sped up the process of medication discovery and development.

Artificial Neural Networks, brain-inspired computer architectures created to mimic how the human brain learns, are the basis of artificial intelligence. With the ability to learn for themselves, artificial neural networks can improve their output as additional data becomes available. Therefore, artificial intelligence enables a computer to complete tasks that would normally require human intelligence either autonomously or with minimal human involvement. To put it another way, machine learning algorithms pick up information from the programmer's inputs and use it to generate new data by independently making judgments. As a result, the programmer sets the parameters while the AI creates the task. The majority of AI applications, from chess-playing machines to self-driving automobiles, can be observed to strongly rely on deep learning and natural language processing. With the aid of these technologies, computers may be taught new skills to carry out certain specialised activities, such as producing creative material by analysing vast amounts of data and identifying certain particular patterns in the information provided to them.

Intersecting Artificial Intelligence with Intellectual Property Rights

Artificial intelligence is used widely today. Humans cannot currently avoid living without technology in any sector of the economy. Whether it be in science or sports, AI is becoming a necessity in everyone's life. Artificial intelligence (AI) is a helpful tool in the subject of IP. Many industries and organisations that value contemporary software applications use it. Patents help with AI invention regulation and protection. WIPO has promised that IP laws will start to acknowledge the need for AI techniques to be implemented and developed in every business. Both AI and IP involve the use of technological advancements that are controlled by enormous organisations in order to conduct business effectively.

IP rules could only protect advancements in AI and issues with legality while also having a significant positive impact on the economy, the creation of goods and services, and their distribution. Therefore, since both AI and IP deal with inventions involving patents, trademarks, and copyrights, they have overlapping objectives because important aims of IP policy include fostering innovation and discovery in technological domains. Because patents are a type of intellectual property that must be protected, the requirement to get patents for AI software developers' technologies draws attention to IP once more.

Through its conferences and regulations, WIPO has long emphasised the importance of AI and how it may be protected by the IP sector as well as incorporated into business strategy to avoid concerns like patent or copyright encroachments. Machine learning in the form of online data is built on information, which would be the core of property rights and innovation in a globalising economy. To keep the rules open and provide legal answers for AI inventions, IP laws must be linked with AI and advances must be regularly tracked.

Technological Trends and Artificial Intelligence

According to the 2019 WIPO report on AI technology trends based on PCT filings, forty percent of AI patent applications involve inventions using machine learning, notably deep learning and neural network techniques. Additionally, it reveals that image recognition is a component of computer vision, which accounts for close to 50% of functional application patent filings. The other two most frequent functional implementations are voice and natural language processing, with robotics and production control supposedly on the rise.

Transportation, including avionics, autonomous cars, and aerospace, telecommunications, including internet, broadcasting, and communications, and life sciences, including biosciences, neurosciences, and medical technologies, are the report's top AI-applied industries. The USPTO's (US Patent and Trademark Office) 2020 report, "Inventing AI," charts the spread of AI across several technologies. The report contains patent information that was discovered by a machine learning system. According to the report, between 2002 and 2018, the number of applications for AI-related patents surged by 100%, going from 30, 000 to 60, 000 every year. According to the research, between 1976 and 2018, their share in various technology sub-fields increased from 9 to 42 percent, and their proportion of all patent applications increased from 9 to 16 percent, demonstrating a considerable increase in the dissemination of AI into various areas. According to US patent applications, the knowledge processing and planning/control innovations in AI are expanding at the highest rate. These innovations are followed by AI hardware, computer vision, machine learning, and evolutionary control.

India ranks higher in scientific publications than in patent filings, according to the WIPO study on AI technology trends. According to the research, the number of AI patent applications with an initial Indian file increased by 33% between 2012 and 2015. India ranks fifth in distributed AI applications and has high-ranking publications in the fields of fuzzy logic and machine learning. Thus it is evident that there are numerous evolutionary tiers of AI that function with little to no human interaction or that are entirely independent. Furthermore, it is anticipated that independent AI will create an exponentially greater volume and variety of material and creativity in the future. However, there are few options for getting IP protection for this kind of invention or work. Given that AI systems function autonomously, it is challenging to identify a single actual inventor. The issue of who will hold the intellectual property rights to whatever inventions it produces arises as a result.

Role of Judicial Agency

The ethical issue of credit of the new evolution done by the artificially intelligent machine was identified and raised in various judgements which are discussed as under:

Ownership of AI related inventions: Dabus Case

Many firms have been compelled to re-evaluate their tenets and embrace a more progressive corporate strategy that incorporates AI innovation as a result of artificial intelligence (AI). When AI takes on the role of creator or author, the more serious problem surfaces. The aim is to address the larger question of whether an AI may be considered an inventor in light of the Dabus case, as well as how the Dabus case was an opportunity lost to create jurisprudence on the subject.

The most recent well-known case to come out of the United Kingdom is Stephen L. Thaler v. Comptroller General of Patents, Designs, and TradeMarks. Two patent applications for patent grants, GB1816909.4 and GB1818161.0, were submitted in Thaler's name to the United Kingdom Intellectual Property Office ("IPO"). Since Section 30 of the Patents Act of 1977 indicates that the right to apply for a patent is transferable, it is possible that Thaler was not an inventor as stated in the application. The IPO then informed Thaler that she needed to submit Patent Form 7, a statement of inventorship and the right to grant patents, in line with Section 13 of the Act. Thaler submitted Patent Form 7 in which he claimed that the Device for the Autonomous Bootstrapping of Unified Sentience (or "DABUS ") AI is the true inventor and that Thaler has the right to grant patents as the AI's owner. Or, to put it another way, Thaler owns Dabus and Dabus is the inventor. Due to Thaler's ownership of Dabus, Dabus has given Thaler the authority to grant patents. The reasoning therefore assumes that Dabus has a right to patents and can afterwards transfer those rights to its owner, Thaler. The topic of whether or not an AI can receive a patent has been raised as a result of this fundamental premise. The Special Patents Court of the High Court of England and Wales overruled the IPO's decision and upheld it.

The UK Copyright, Design and Patents Act, 1988 states that any literary, dramatic, musical, or artistic work that was computer-generated must have all essential arrangements made by the author; however, it does not address patents and innovations. According to Section 306103 of the Compendium of the US Copyright Office, registration of authorship will only be carried out if the work was created by a human, leaving no room for expressions produced by artificial intelligence, much alone for copyrights. The term "person" or, in the case of a joint invention, "individuals" is all that is meant by the term "inventor" under US law. Therefore, by using the term "individual", the US excludes corporations as well.

Both the IPO and the Special Patents Court used the same justifications to reach an agreement. Dabus cannot hold Act-related patents because it is a computer and is not a recognised legal body. The identity of the inventor was investigated by the court. The answer, though, seems to be really simple. The Court repeatedly said that Dabus cannot be granted the patent because it is not a "person", without going into the literal understanding of what a "person" can be. There doesn't seem to be a consensus on what makes someone a "person". A person is a natural being as well as a legal creature. As a component of a legal entity, corporations are created. The denial of a patent to Dabus may have profound policy repercussions that affect the entire planet. Who will own AI inventions has been a topic of renewed debate as a result of the reluctance to grant patents. As the author has already mentioned, this case was a missed chance to create a regulation that would, at the very least, give the innovators some explanation. The author thinks that the court might have established some guidelines that would have provided guidance in the future for similar circumstances. The court relied on the argument that as AIs are not people, they cannot be issued patents, and thus left the topic of who should own AI creations to a more general policy debate.

Burrow Giles Lithographic Co. v. Sarony

The major issue in this situation is whether or not a photograph could be protected by copyright. It was important that the court in this case recognised the difference between creative and mechanical labour. In this ruling, the Court reviewed whether to provide copyright protection to the output of the machine and came to the conclusion that mechanical output is not creative in and of itself. As a result, it will be challenging to secure the intellectual property of the work that AI technologies produce if such a rigid approach is taken with them.

Bleistein v. Donaldson Lithographing Co.

In this instance, the court recognised a distinct difference between human effort and machine activity. In this judgement, Justice Holmes and the majority emphasised the significance of determining a person's eligibility for a copyright and urged that it be acknowledged as the most crucial factor in doing so. The phrase "Something Irreducible" used by the court to describe its position meant that no copyright could be assigned to something that was not made by humans.

Alfred Bell & Co. v. Catalda Fine Arts

In this instance, the court treated copyright distribution more kindly. The judge decided that in order for a piece of art to be considered original, it should not be completely reproduced or plagiarised from earlier works of art. Even more, the court decided that an author might claim ownership of any unintended change. Because it is not copied as a result of this decision, the output of the AI system is entitled to copyright protection. This makes clear the position's ambiguity with regard to the protection of AI system works. The rights of holders are impacted, though, by the lack of a firm strategy.

Ferid Allani V. UOI

In this instance, a device and access method for web-based information sources and services are being patented. The Controller rejected the patent request. Based on the fact that only computer programmes per se are not patentable in India and that a computer programme that has a technical effect or contributes in a technical way is not a computer programme per se, the Court ordered the Controller to re-examine the patent application. The Court ruled in the case that computer programmes integrated into digital or electronic devices must be tested for technical efficacy and that inventions involving computers must be evaluated in light of available technology.

It asserted that computer programmes cannot be denied patent protection if they comply with the legal standards. The court ruled that it would be outdated to claim that all inventions in the modern digital age, in which the majority are based on computer programmes, would not be patentable. Computer programmes would serve as the foundation for innovation in the fields of artificial intelligence, block chain technology, and other digital goods, but they would not qualify as patentable innovations.

The recent decision appears to support long-standing CRI legislation, which stipulates that any innovation having a technological component or impact is not merely a software programme, even if it is carried out with the use of a computer programme.

Indian Scenario

Domestically, the Tencent case supports the notion established under Section 2(d)(vi) of the Indian Copyright Act of 1957, which awards authorship to the person who "causes it to be formed". Indian courts have in the past emphasised the importance of human interaction in the creation of copyrighted works. Additionally, in some jurisdictions, "Decisive" human involvement is acknowledged as a prerequisite for granting copyright protection. A creative team has been hired by the plaintiff in the Tencent lawsuit to oversee the use of the AI software.

In the case of Naruto v. Slater, Due to the fact that only a "natural person" could be listed as an inventor, the EPO declined to award the patent. It appears that human interaction offers a workable approach that enables IP offices to maintain their core concepts of "Authorship" and "Copyright". The appropriate degree of human involvement, however, is widely contested. Similarly, the Indian Copyright Office views decisive human involvement in AI-created works as a necessary condition for granting IP rights.

As a result, a strict reading of section 2(d) reveals that while a "person" who causes a work to be created in India shall be accorded copyright, the same is not true of computer-generated works across the board. This may only apply to weak AI that needs human engagement, but it is unconscionable with strong AI, which may not need human involvement at all. The three separate stages of AI are intriguing to explore in light of the aforementioned are:

- Artificial Narrow Intelligence,
- Artificial General Intelligence, and
- Artificial Super Intelligence.

The first step, which is often seen in modern tech gadgets like voice recognition software, manufacturing robots, and even self-driving automobiles, refers to traditional weak AI. This kind of artificial intelligence mainly depends on people to programme it to carry out specific tasks. Strong AI, in contrast, describes devices with "human-like" intelligence that can be used to solve issues. Even further away is the development of artificial superintelligence. The Tencent case decision seems well-cut out that such AI programmes can be deemed to fall neatly under section 2(d), given that the development of AI rests, at best, on the spectrum between a weak and an almost powerful AI (vi). This could change, though, as technology continues to advance toward the creation of purely powerful AI. So the question is, will people involved in an exclusively AI-created work be given authorship, or will there be no authorship at all? By making reference to section 17 of the Indian Copyright Act, 1957, the aforementioned query is answered in the Indian context. Given the limited definition of "author" in India, it is extremely doubtful that an AI machine would be included under sections 17(a), (b), or (c). However, as AI-created works are unlikely to fall under the definitions of sections 17(a) or (b) and since an AI-machine is not acting in the course of employment of a business or an individual, they may still not qualify as works by authors. The same is true for the claim that an AI machine can complete the requirements of a section 17 service contract (c).

As it stands, section 2(d)(vi) of the statute gives authorship to the person who "causes the work to be made" in the case of computer programmes with weak AI. However, where there is no human interaction, it is impossible to claim authorship of the final product. Similar to that, AI-created works are not exempt from the principles outlined in section 17 of the aforementioned Act.

Conclusion

Claims that AI technologies ought to have their own legal framework are gaining support as they get a little more advanced and get more attention from the public. Often, the arguments have been presented in utilitarian terms and compared to other legal organisations, such corporations. The implication of these claims, or the explicitness of one's knowledge and examples, is that AI systems should be given the same status as actual people when they become indistinguishable from humans.

Although the inclusion of AI-based works in the copyright legal framework raises important policy questions, a different possibility has already emerged in which AI-generated works may not meet the criteria for originality. Rightfully, current regimes prioritise the preservation of human creations over AI-based works, yet it may also be crucial to take into account finding a fair balance between the advancement of AI technologies and human creativity.

Recognizing that IP rights may go beyond the usual ideas of IP laws is one method to achieve this. In that regard, it might be advantageous for governments to compromise their legal stances in order to embrace a new definition of "authorship" that takes AI into account. At the same time, it might also be challenging to disregard the possibility that "AI-created" works produced without human input would not give rise to authorship claims on their own. First, we would like to point out that there can never be a claim of "no human involvement", as copyright law only allows for specific claims of authorship.

References

John Haugeland, ARTIFICIAL INTELLIGENCE: THE VERY IDEA, 1st ed. 1989, Bradford Books.

Klaus Henning, HOW ARTIFICIAL INTELLIGENCE IS TRANSFORMING THE WORLD, 1 st ed. 2021, Springer publications.

V.K. Ahuja, LAW OF COPYRIGHT AND NEIGHBOURING RIGHTS NATIONAL AND INTERNATIONAL PERSPECTIVES, 1st ed. 2015, Lexis Nexis.

Artificial Intelligence, WIPO Technology trends, available at https://www.wipo.int/publications/en/details.jsp ?id=4386 (Accessed on June 10, 2022).

Ryan Abbott, THE REASONABLE ROBOT: ARTIFICIAL INTELLIGENCE AND THE LAW, 1 st ed. 2020, Cambridge University Press.

Available at https://www.asiaiplaw.com/section/in-depth/ai-intersects-indias-patent- regime (Accessed on June 2, 2022).

Woodrow Barfield, ADVANCED INTRODUCTION TO LAW AND ARTIFICIAL INTELLIGENCE, 1 st ed. 2020, Edward Elgar Publishing Ltd

Klaus Henning, HOW ARTIFICIAL INTELLIGENCE IS TRANSFORMING THE WORLD, 1st ed. 2021, Springer publications.

Woodrow Barfield, ADVANCED INTRODUCTION TO LAW AND ARTIFICIAL INTELLIGENCE, 1st ed. 2020, Edward Elgar Publishing Ltd.

Hikari Senju, How AI Regulations are Shaping Its Current And Future Use, Forbes, available on www.Forbes. com/sites/forbestechcouncil/2022/07/14/how-ai-regula tions-are-shaping-its-current-and-future-use/?sh=2a c689983d40 (Accessed on June 12, 2022).

Ryan Abbott, THE REASONABLE ROBOT: ARTIFICIAL INTELLIGENCE AND THE LAW, 1st ed. 2020, Cambridge University Press.

Klaus Henning, HOW ARTIFICIAL INTELLIGENCE IS TRANSFORMING THE WORLD, 1 st ed. 2021, Springer publications.

Available at https://www.wipo.int/edocs/pubdocs/en/wipo_pub_1055.pdf (Accessed on 02 July.2021).

Kalayan C. Kankanala, Artificial Intelligence (AI) Inventions and Patents, BananaIP, July 24, 2021.

V.K. Ahuja, LAW RELATING TO INTELLECTUAL PROPERTY RIGHTS, 1 st ed. 2017, Lexis Nexis.

V.K. Ahuja, LAW OF COPYRIGHT AND NEIGHBOURING RIGHTS NATIONAL AND INTERNATIONAL PERSPECTIVES, 1 st ed. 2015, Lexis Nexis.

Ryan Abbott, THE REASONABLE ROBOT: ARTIFICIAL INTELLIGENCE AND THE LAW, 1st ed. 2020, Cambridge University Press.

❑❑❑

8

E-Commerce and Challenges in Online Arbitration

Dr. Sanya Yadav
Assistant Professor, School of Law, Bennett University

Abstract

Due to the unlimited jurisdiction and active operations of electronic business, the parties to any transactions related to the e-commerce sector are governed by contractual obligations under an e-commerce contract held between them. This implies that, in case of e-commerce transactions between two parties or two countries of different legal fraternity, the arbitration rules apply and these rules as firmed by the parties themselves. Therefore, if any breach of obligation happens in the contract between the parties that will arise dispute. Primarily, parties initiate to resolve the dispute themselves through consultation depending on the circumstance. If parties are not satisfied with the consultation, then parties can go for the Alternative Dispute Resolution Mechanism (hereinafter ADR) for settlement of disputes. This practice, is to be conducted in pursuance of the international law rules which has been accepted by the legal system of such parties.

Presently, many difficulties arises with respect to adequate understanding of rules and laws for the settlement of disputes by arbitration tribunals. This research will aim at understanding the multifaceted choices available to the parties to establish and facilitate in e-commerce and also covers the various efforts of online arbitration towards forming the uniform legal framework among its member countries to expedite the same. Online arbitration is the best designed and sophisticated platform to increase e-commerce and their dispute settlement process. It is also very cost effective and works for the best interest of the parties. Yes, many individuals have been backed away from the courts due to unwanted delays, but on the day everyone accept it as an convenient and alternative mode of online resolution for the settlement of disputes in e-commerce, no one will be going back from choosing this way to for dispute settlement in arbitration.

Keywords: *E-Commerce, Online Arbitration, Alternative Dispute Settlement, Arbitration*

Introduction

Over the past two decade with the advent of twenty first century, the international communication and trading style has been changed and broadened the scope of cyberspace significantly and engaged a major alteration in the practice of business over e commerce for reason being reduced supply times and cost in the situation of internal and international trade transactions. These drastic transformation within the ever-changing paradigm of revolutionary practice in trade demands well advanced swift information and blue-prints so as to muddle up with the dynamically emerging trade practices across member states of United Nation (hereinafter UN) and other states, increasing from the inflated use of technology in trade through e-commerce and Electronic Data Interchange (hereinafter EDI).

EDI and e-commerce terms are perplex in nature and not similar in nature. E-commerce is a generic term which encompasses electronic communication technologies as well as other EDI, such as internet and electronic mail. It can thus be defined as any form of administrative transaction or communications technology or business or business information exchange which is executed using information.

In recent years, technological developments are deliberately developing customary arbitral procedures along with arbitral practices. The future of online arbitration is based upon the submissions of emails and online video conferences. As a new dimension, the legal literature has already defined the vision of new technology to encounter the regime of International Commercial Arbitration (hereinafter ICA). Arbitration encompasses many forms of different online methods for the practice in this era. Arbitrators across the globe want to measure without departing their hometowns and would willingly issue an arbitral award in an electronic form. The purpose of this research is to discuss how such high-tech revolutions can be improved by the current regime of ICA and legislative framework.

As the drastically advent of e-commerce transactions, it is predictable that the e-commerce dispute will also rise in the society. It can be raised due to many reasons such as unsecured websites, defective products and fake information. To resolve these kinds of problems from society, consumers must be aware about their constitutional right to access the domestic courts under the Indian Constitution. Consumers should be responsive about enforcement of their rights. The main restraint is the number of backlog cases in the courts, which prevent parties from getting justice. That will lead to finding an alternate way to get the prompt, efficient and cheaper way to resolve the disputes among the parties. To bridge the gap, this research will explore more on the alternate methods like online dispute resolution to provide justice to the consumer in e-commerce.

E-Arbitration Agreements

Instead of traditional form of arbitration, the countries often use the arbitration agreement as modern means of electronic transmission. The legal background, which is more important for such agreements, including the New York Convention (1958), which was highly recognised in many countries before the internet era. Thus, begins with the delinquent of whether it is feasible to truly agree on arbitration determined over the electronic means, such as by taking assent via email. In the present scenario, whether national or international context arbitration is a significant and preferable method for the dispute settlement. The focus should be on the draft of the written arbitration agreement which examines and interprets the situation according to the New York Convention. Most important, the e-contracts must be finalized in writing and duly signed by parties. However, as per Article 1316 of French Code Civil *"Documentary evidence, or evidence in writing result from a sequence of letters characters, figures or of any other signs or symbols having an intelligible meaning, whatever their medium and the ways and means of their transmission may be."* Due to the advent of technology, the mode of communication between the parties for finalizing the contract via internet, they don't conclude the contract on paper. In the international transaction, parties conclude a contract with even single meeting. Therefore, parties often include arbitration in their e-commerce for the settlement of disputes. But sometimes, domestic courts do not recognise the validity of e- arbitration agreement. There is need when parties conclude an e–agreements they should include feasible solution for the disputes in the e-commerce and their terms should acquiescence with the New York Convention.

E-Arbitral Proceedings

Technology is already extensively used in arbitral proceedings. Forms, papers and credentials can be transferred directly to parties and arbitrators at a unexceptional cost with the use of the Internet. Due to the result, various arbitrators appreciated that work is simpler than the hard copies due to electronic transmission of documents, expressly when compliances have abundant sheets. The internet appears to have an additional considerate influence on arbitration practices while using information technology on a daily basis. Customarily, meetings of parties were a necessity in arbitration proceedings and arbitrators preferred to appear in person or through appropriately sanctioned governments. At present, it can be observed that the internet inspires inaccessible mechanisms for the settlement of dispute and corporeal conferences are often eradicated, or precisely, changed through varied electronic discussions. These expansions appear to fetch assistance for both the parties and arbitrators, even with leaving their places the arbitrators can proceed, the parties evade inviting cost of travel.

The Seat of Arbitration

If the proceedings of arbitration entirely conducted online, with arbitrators and parties in different places, *prima facie* it appears difficult or in fact difficult it is to regulate the seat or place of the arbitration. The opinion pointed out by few scholars in their writing "*virtual arbitration has no situs*", in fact "*no identifiable seat of arbitration, is not pertaining to any particular geographical territory*".

Since the procedural acts in online arbitration are performed remotely, "*determining the place of arbitration on the basis of objective indices is scarcely conceivable*". Dispute between the parties allows the arbitration seat to be "*a strictly legal concept dependent on the will of the parties*".

Online Dispute Resolution in India

In India, the pendency of cases before the several courts itself provides necessary purpose to completely consider bringing in some relevant modifications. One of the important measures adopted by few states, by raising the court fees for many types of proceedings which are mandatory to pay while filing the petition in tribunals and courts which is mainly to create barriers or restrictive measures at the initial level.

The Arbitration & Conciliation Act, 1996

Precisely it is difficult to analyse the gaps in the 1996 Act with respect to the provisions relating to conciliation proceedings as against to arbitration proceedings is that the 1996 Act does not impose limits on the application of its provisions to conciliation proceedings taking place in India only. This provision hence affords parties the adaptability to hold their proceedings anywhere, even in cyberspace. This does not, however, mean that arbitration proceedings cannot be held online if they are to be manageable to the 1996 Act, for the said Act acknowledge the autonomy of the party and does not place restrictions on applicable law and their choice of procedure as long as they have all agreed thereon.

According to the provisions of the 1996 Act, after inviting other party to conciliate, parties can start the conciliation proceedings with in 30 days that should be accepted by the other party,. Afterwards, the appointment of conciliators would take place – there can be one, two or three conciliators. By mutual consent parties can either appoint one each or one conciliator, or and they also appoint third if it is requested by the appointed conciliators.

The conciliators shall lead the way by principles of justice, equity, good conscience, objectivity and fairness in their venture to help the disputed parties in an impartial and independent way that shall preserve the outcome as well as the confidentiality of the proceedings. The conciliation recommendations are not binding upon them, if parties feel they can terminate jointly or unilaterally the conciliation proceedings. Nevertheless, if parties sign a settlement agreement which is duly certified by the conciliators then the same will be binding upon the parties and court can enforce via decree.

Settlement of Disputes Outside the Court According to the Provision of The Code of Civil Procedure, 1908 (The Code)

In India, Section 89 of the Code, elaborated the essence of arbitration proceedings. The provision brings out the settlement process outside the court to minimise the cost and burden of the court. A separate legislation was formed in 1940, the Arbitration Act 1940 which dealt with arbitration in India, that invalidated section 89 of the code. Again, after half of the decade a new provision was introduced by the CPC (Amendment) Act, 1999 for rules related to Section 89. It provides procedures which settle the disputes peacefully and amicably between the disputant parties. This section reads as follows:

"Where it appears to the court that there exist elements of a settlement which may be acceptable to the parties, the court shall formulate the terms of settlement and give them to the parties for their observations and after receiving the observation of the parties, the court may reformulate the terms of a possible settlement and refer the same for:

- arbitration;
- conciliation;
- judicial settlement including settlement through Lok Adalat; or
- mediation."

"Where a dispute had been referred-

(a) for arbitration or conciliation, the provisions of the Arbitration and Conciliation Act, 1996 shall apply as if the proceedings for arbitration or conciliation were referred for settlement under the provisions of that Act.

(b) to Lok Adalat, the court shall refer the same to the Lok Adalat in accordance with the provisions of sub-section (1) of section 20 of the Legal Services Authority Act, 1987 and all other provisions of that Act shall apply in respect of the dispute so referred to the Lok Adalat;

(c) for judicial settlement, the court shall refer the same to a suitable institution or person and such institution or person shall be deemed to be a Lok Adalat and all the provisions of the Legal Services Authority Act, 1987 shall apply as if the dispute were referred to a Lok Adalat under the provisions of that Act;

(d) for mediation, the court shall affect a compromise between the parties and shall follow such procedure as may be prescribed."

Rules and Regulations under Information Technology Act, 2000

According to the Information Technology Act, 2000 Act (the Act) the parties must encourage themselves for e-commerce in the society and provides legal recognition to digital signatures and electronic records. The advent of this legislation has also taken up many modifications to other legislations for instance, The Indian Penal code 1860, The Reserve bank of India Act 1934 etc. It is a law meant to be "*applicable to alternatives to paper based methods of communication and storage of information.*"

According to section 4, "*states that a requirement of any law for information or matter to be in printed, written or typewritten form shall be deemed to have been contended if such information or matter is rendered or made available in an electronic form and is accessible so as to be usable for a subsequent reference."* As per section 5 elaborate *"recognition of digital signatures by providing that the requirement of any law for authentication by a person's signature shall be deemed to have been satisfied if such authentication is done by means of a digital signature."*

Implementation of Online Alternate Dispute Resolution

From the last two decades, The Supreme Court of India has approved the use of technology for the settlement of disputes. In a case it was held that for the purpose of taking evidence of a witness video-conferencing could be utilized. Also, explaining the benefits of video-conferencing, the Court stated that *"In fact the Accused may be able to see the witness better than he may have been able to if he was sitting in the dock in a crowded Courtroom. They can observe his or her demeanour. In fact the facility to play back would enable better observation of demeanour. They can hear and rehear the deposition of the witness."* Many admissions have been made and the Court mentioned that *"Virtual reality is a state where one is made to feel, hear or imagine what does not really exist. Video-conferencing has nothing to do with virtual reality. Video-conferencing is an advancement in science and technology which permits one to see, hear and talk with someone far away, with the same facility and ease as if he is present before you i.e. in your presence." "This is not virtual reality, it is actual reality."* Many fruitful observations are made in the said order which clearly indicate the strong support for use of technology specifically video-conferencing in the justice delivery system which is highly adopted in India during covid time duration.

According to the precedents, not for the first time the Supreme Court has marked in favour of information technology. Previously also decision came and it held that "When an effective consultation can be achieved by resort to electronic media and remote conferencing it is not necessary that the two persons required to act in consultation with each other must necessarily sit together at one place unless it is the requirement of law or of the ruling contract between the parties." In this situation, the contention was that for the appointment of the third arbitrator the appointment of two arbitrators should be done by the parties themselves.

Following the above judgement the another judgement in which use of available technology has been given a real boost, the Supreme Court held that "Technological advancement like facsimile, Internet, e-mail, etc. were in swift progress even before the Bill for the

Amendment Act was discussed by Parliament. So when Parliament contemplated notice in writing to be given we cannot overlook the fact that Parliament was aware of modern devices and equipment already in vogue."

Conclusion

The tradition of online arbitration is developing which helps in the growing awareness in the society. Few questions have been asked in legal discussions regarding, validity of virtual arbitration in the present legal framework. As an upcoming mechanism for the settlement of disputes, online arbitration has confronted many problems while applying the customary principles of arbitration law in commercial transactions or any investment transactions.

This research highlights the latent difficulties which are coming in the way of online arbitration. In international arbitration, starting problems raise to formal requirements in the present situation. These requirements come in various forms: such as signatures requirement, the originality and requirement of an agreement in writing. The disagreement is going on, under few circumstances online arbitration should not be troubled by recognised desires. In conducting arbitral proceedings, the advanced technology may be of great assistance, including hearings, negotiations between arbitrators and electronic submissions.

In this research few issues have been also quickly summarised such as the place where the award is made and describing the seat of arbitration, subsequently, in '*cyber-arbitration tribunals*' there is physical seat therefore online arbitrations might be far better. It is established with the view that in the international arbitration law the expansion of online arbitrations may provide a new meaning to the theory of delocalization.

It is fundamentally hard to reduce the number of disputes in e-commerce through certain setbacks, and thus further online arbitration *is* not accessible as an actual substitute to customary arbitration. Nevertheless, it is reflected as a legal instrument that is capable however "*must be utilized with caution due to the above-mentioned legal uncertainties.*" Online arbitration provisions determine an innovative aspect of customary arbitration, prevailing in the identical legal framework. In this view, beginning legal difficulties do not establish impossible difficulties to the submission and promote the progress of online arbitration. The increase of sources in online arbitration appears to replicate this opinion. Justice is taking too long a time, not available for the people and expensive for the individual that is not considered as justice. What is lacking is not only awareness of this moment amongst the people but also the expertise which is necessary to execute online arbitration as a truly feasible and much easier substitute method to challenge in a court of law.

References

Lynch K., (2003). *The Forces of Economic Globalization: Challenges to the Regime of International Commercial Arbitration.*The Hague: Kluwer Law International.345.

Katsh, E. E., Katsh, M. E., & Rifkin, J. (2001). *Online dispute resolution: Resolving conflicts in cyberspace.*

State of Maharashtra v Dr. Praful B. Desai (2003) 4 SCC 601(India)

Grid Corporation of Orissa Ltd. v AES Corporation (2002) A.I.R. SCC 3435(India)

Sil Import, USA v Exim Aides Exporters, Bangalore (1999) 4 SCC 567

❑❑❑

SECTION - 3
E-Commerce Sector

9

Digital Payments in Digital India: A Discourse

Mudit Jain
Research Scholar, School of Management & Commerce, Manav Rachna University

Chahat Wadhwa
Assistant Professor, Fairfield Institute of Management and Technology
(Affiliated to Guru Gobind Singh Indraprastha University)

Sonam
Assistant Professor, Fairfield Institute of Management and Technology
(Affiliated to Guru Gobind Singh Indraprastha University)

Abstract

With the demonetization coupled with Covid-19 Pandemic, dependency of Indians on digital payments has increased. If not fully, digital payments have replaced traditional cash payment mechanisms in almost all commercial transactions. Use of digital payments opened the door for the Digital Indian Economy, allowing digital transactions to take place in every economic activity. It poured certain advantages in the day to day commercial activities of Indians. It promotes instant payment, saves time, builds a customer base and promotes transparency and accountability with respect to commercial transactions. However, just like every coin has two sides, digital payments have their own pros and cons. For instance, digital payments suffers from challenges in the form of infrastructural issues, financial cum digital illiteracy, hacking, lack of public trust etc. The present paper discusses the usefulness and drawbacks associated with the use of digital payments in commercial transactions in the Indian Economy along with suggestions towards making digital payments mechanisms more effective and secure. Key suggestions include promoting digital literacy through seminars and social media platforms like YouTube, Facebook etc. , formulating a stringent legal regime for curbing cybercrimes as well as incentivizing digital payments through concession and rebates etc. The researchers have adopted a doctrinal method using exploratory research for collecting data primarily from secondary sources.

Keywords: *Digital Payments, Digital Finance, Commercial transaction, Unified Payment Interface (UPI) and Cybercrime*

Introduction

The 21st Century can also be called the century of digitalization. Technology has become a part of today's society, touching every sphere of life of an individual including the economic sphere. In today's world not only the mode of offering, buying, selling and promotion of goods and services have become digital but also the mode of payment for buying such goods and availing such services has also become digital. World is witnessing a gradual shift from cash based economy to digital economy wherein transactions are done electronically at the click of a button either by using digital payment apps like Paytm or by bank cards. It allows entrepreneurs to make electronic payment of goods and services via mobile phones, internet and other available access points other than using cash or cheques.

Digital payment mechanisms increase participation of entrepreneurs in e-commerce transactions, curtailing operational costs, easier management of trade contracts and delivery records. Furthermore, it helps in improving their relationship with clients, vendors, financial institutions like banks and government.

Digital payments involve cashless transactions which improves environmental sustainability as no trees are cut down for making paper used in printing currency notes, demand drafts etc. The reduced production, printing and distribution of currency notes, leads to lowering down carbon footprints thereby, protecting the environment. Furthermore, digital payments are more hygienic in nature than cash payments as there is no or very less chances of contamination or transferring of diseases as, no physical contact either directly or indirectly is involved in digital payments.

Growing Reliance on Digital Payments in Indian Economy

Over the course of recorded history, the consensus regarding the origin and forms of ancient money has shifted. However, the function of money remains the same, facilitating the movement of goods as a medium of exchange and store of value. The form in which money has been used over a period of time has changed from time to time on the basis of convenience. Since the beginning of Human civilization money was used in the form of a coin replacing barter system, the rulers used to produce and circulate the coins within their kingdom in their names and images. With the invention of printing press currency notes of paper form of money came into existence which slowly replaced the role of coin as medium of exchange in commercial transactions. Now, in the present era of digitalization this role has been shifted to Digital payments. It is also pertinent to note that any change in the form in which money has been used over the years is completely based upon the convenience of the people.

The recent policy initiatives and technological advancements have propelled India's Digital Payment mechanism. Talking about the contemporary world, growing technology coupled with globalization has not only made the world well connected as a single family but has also led to increased commercial transactions. The volume, nature and kind of the goods being traded today are far more bulky and vibrant in nature, than earlier. Such a huge amount of commercial transactions requires efficient settlement of payments to be made over long distances with minimum transaction cost. Therefore in order to meet this requirement digital form of payment mechanism is needed. Payment via Cash is of course ingrained in our habits and being easily accepted everywhere, still remains a medium of exchange. However, payment through digital platforms is far more convenient, reliable, saves time etc..

The Indian economy is based upon two sectors i.e. organized and unorganized sector. Organized sector is largely regularized and commercial transactions in the organized sector can be monitored easily. While the unorganized sector is largely non-regularized, involves black money circulation and employs the majority of the Indian population. Digital payments used in commercial dealings in both sectors allows government and concerned authorities to keep a check on these transactions as such transactions involve facilities of banking institutions. This enables the government to monitor commercial transactions taking place, tracing the chances of involvement of black money in such transactions, detecting tax evasion and any unlawful transactions from taking place which might not be possible if such transactions involved use of cash. Furthermore, in recent years the progress in digitization has been fueled by the infrastructural support such as mobile applications, government policies and regulations. The Indian Government and RBI has introduced schemes like Pradhan Mantri Jan Dhan Yojana,

Aadhar card, Kissan Credit Card, Atal Pension Plan etc. to motivate the public in becoming part of the banking system. Further, policies and laws like Demonetization and Goods and Services Tax (GST) have further brought the public closer to banks. Introduction of Payment systems like NEFT (National Electronic Funds Transfer), Unified Payment Interface (UPI), Real Time Gross Settlement (RTGS), IMPS, cards and Pre Payment Instruments have increased the consumers' payment options. Furthermore, Introduction of Point of Sale (PoS) Terminals along with infrastructural support in the form of QR code has boosted digital payments mechanisms.

Introduction of the National toll Collection System (NETC), Bharat Bill Payment System (BBPS) and RuPay cards have boosted the digital payment system in the nation. The Government's intention to incorporate modern-day technologies such as tokenization and contactless payments will accelerate the process. The National Highway Authority of India (NHAI) has made it mandatory for the vehicles using the highway to get FASTag (a digital payment mode on any of the toll plazas) for the purpose of paying toll tax.

Digital payments have bestowed benefits to the customers by offering a variety of services at the click of a button. For instance, the world's largest search engine Google started Google Pay which is a B2C business which not only turned out to be successful but transformed UPI based payment mechanism globally. It provides services such as bill payments, mobile recharges, food and lodging etc. This makes it easier for the customers in terms of saving precious time, being cost effective and provides a large amount of choices.

This shows there has been a growing reliance on digital payments in the Indian economy and it may slowly bypass the use of cash in making commercial transactions.

From the above discussion it is clear that digital payments has a bright and illuminating future in Indian Economy, bestowing various benefits to Indian citizens such as:-

- **Time Saving:** Cash transactions often involve physical movement of person and individuals. Sometimes individuals have to travel far for making payments in commercial dealing especially where huge amounts of money is involved. Individuals do not have to wait as online payment mechanisms have eliminated the need to visit banks either for withdrawal of cash or presentation of cheques involved in commercial transactions. Unlike in case of cash or cheque, one does not have to worry about time and location while making online payments as the same can be done anywhere and anytime. This prevents wastage of time in making commercial transactions i.e. no need for standing in long queues outside banks as one can easily make payments via mobile apps.

- **Instant Real Time Payment:** Online payments helps transfer of funds instantaneously 24*7*365 through debit cards,credit cards,E-wallets,UPI, QR code ,mobile apps or bank accounts directly linked with registered mobile numbers. Cash payments to a far-off vendor or supplier may take a long time, while digital payments can be performed quickly in a few seconds. The process of clearing cheques makes them another time-consuming method of payment.
- **Provides security in commercial transactions:** Digital payments reduces the risk of loss, theft and other financial crimes as being faced in cash based transactions. Further, in order to make digital payments, certain security checks are being created prior to making digital payments online such as password protection, face and fingerprint recognition and use of One Time Password (OTP). Such checks provide security to a certain extent in transactions involving digital payments.
- **Ease in Record keeping and Tracking:** Digital Payments enables individuals to keep a track on the transaction entered by them as records of all transactions are maintained irrespective how small the amount would be Small business owners can monitor their sales by category of products and day of the week and utilize this data to improve inventory control.
- **Develop customer base:** Customers are more likely to use electronic payments now that they are being rewarded with cashbacks and discounts, which has further contributed to the development and expansion of supplier's client base. By doing this, a firm can turn its current customers into brand ambassadors, boosting its reputation and generating higher profit margin.
- **Promotes Transparency and Accountability:** Transparency and Accountability are perhaps one of the critical issues in commercial transactions. Traditional payment mode involving cash and cheque suffers from the issue of transparency and accountability. There are chances of involvement of black money in commercial transactions and fake cheques may be made for making payments in the commercial transactions leading to the issues like corruption, illegal activities, tax

evasion and cheque bouncing etc. Furthermore, lack of transparency makes it difficult for the concerned authorities to keep a check on such transactions and the crimes conducted ancillary to such transactions. However, if in case any such crime is being detected then another issue may arise with respect to who shall be made accountable for such a transaction? While making digital payments one need not to be worried about the details of the payer and payee as almost all digital payments are linked with banks. This makes digital payment mechanisms more transparent than traditional cash based mechanisms. The digital payments allows the concerned authorities to keep a track on the commercial transactions undertaken by an individual thus, preventing the chances of tax evasions and black money etc.

- **Promotes Green Economy:** Digital Payments reduces the need for felling trees for paper and lowers the price of printing paper currency. Further, one need not have to travel far all the way to banks and financial institutions which reduces reliance or dependence on vehicles for such commutation. This reduced dependence on vehicles reduces vehicular pollution, carbon emissions etc. and helps in saving the environment. Owing to digital payments one need not have to leave the house for fulfilling his financial requirements as the same can be catered by digital finance like Paytm , BHIM UPI, NEFT etc. A single click may be used to complete them instantly from anywhere.
- **Hygienic Commercial Transactions:** Covid-19 pandemic compelled people to find new avenues to make commercial transactions including payments touch free thereby avoiding any human contact. The use of Point of Sale (POS) terminals in business models provides contactless human touch. Under this system one needs to hold his phone near the terminal and payment gets processed automatically. Individuals can also make payments by using QR codes and one time password.
- **Improved financial inclusion:** Prior to the digital payment revolution, access to banks and financial institutions remained limited. Most of the poor remained expelled from the banking services. However, the digital revolution coupled with digital financial services inculcated millions of poor into the banking system. In the words of Jim Yong Kim, former President of World Bank *Universal access to financial services is within reach—thanks to new technologies, transformative*

business models and ambitious reforms... As early as 2020, such instruments as e-money accounts, along with debit cards and low-cost regular bank accounts, can significantly increase financial access for those who are now excluded.". Mobile phone applications and digital technologies provided underprivileged people with services involving payments, transfers, insurance and credit facilities at the click of a button. The Government of India in order to promote inclusion of every citizen of the nation within the banking system has launched initiatives like Pradhan Mantri Jan Dhan Yojna, Kisan Credit Card etc.

- **Facilitates easy access to credit:-** Finance has always remained a big issue for small merchants, farmers, poor and downtrodden class of society. Access to banks and financial institutions still remains a dream for millions living in remote areas of the nation. Further, Demonetization has unveiled the crisis of ATMs in the backward areas of the country. Fintech companies in order to tackle such issues came up with digital finance involving Digital Payments via mobile apps with credit facilities enabling marginalized communities to have an access to credit. Government of India introduced Kisan Credit Card scheme in 1998 with the view to provide sufficient and timely credit assistance from the banking system to the destitute and needy farmers for cultivation including the investment needs of them under a single window with flexible and simplified procedure. Electronic Kisssan Credit Card was introduced with the view to facilitate the Kissan Credit Card Scheme. Banks are given discretion to adopt concerned guidelines as per their suitability with regard to providing Kissan Credit Card facilities. The scheme has benefitted millions of marginalized farmers in meeting credit needs on easy terms for cultivation and other activities.
- **Ensures smooth compliance:** E-payments has helped entrepreneurs in e-filing of returns, payment of taxes, license fees, registration charges and duties leading to smooth and timely compliance and lower cost of transactions. It minimizes travel time and in- person interaction with tax officials. Additionally, it may promote company formalization and broaden the tax base leading to higher revenues for the government.
- **Promotes Women Empowerment:** Digital payments have empowered women in several ways.Women often find it

difficult to travel to distant suppliers or bank offices due to social standards, long travel distances, and family obligations. They can overcome these mobility restrictions thanks to digital payments, which makes it simpler for them to access money and the market. They can make their independent choice as to how the money would be spent while purchasing any product from the market and pay for the same without taking any permission from anyone for the required payment. It is also believed that women's earnings are associated with higher improvement in children's health and higher expenditures on nourishing food, healthcare, and housing.

Challenges Associated with Digital Payments

From the above discussion it is evident that Digital Payments bestows various benefits on the Economy of any nation. However, nothing comes easily as everything has its own cost. Same goes for digital payments. The following are some challenges involving Digital Payments:-

- **Infrastructural Challenges:** Digital Payments no doubt has a bright future in the Indian Economy. However, the reach of Digital Payments facilities is limited primarily due to lack of proper infrastructure. India being a developing nation still suffers from basic issues like frequent power cuts, limited bandwidth, connectivity especially in Rural and far flung areas of the nation.
- **Illiteracy Including Financial Illiteracy:** An individual must have basic awareness of what digital payments are and how they work in order to reap their benefits. Using such digital means of payment is a nightmare for them in a country like India where the bulk of the populace is uneducated and unable to read or write. Additionally, even individuals who are literate have trouble using these apps to access and make payments.For instance, Shivam Shukla a 26 year old boy when asked by a fraudster, who pretended to be an Paytm employee provided remote access to him via screen sharing app. It allowed the fraudster to see every activity on shukla`s device. When Shukla made payment of Rs 10 at the request of the fraudster, it revealed his UPI pin along with the OTP messages for confirming the transactions. The fraudster used credentials for transfering Rs 38,989/- from Shukla`s account. Thereby, causing Shukla to suffer from financial and monetary loss.

- **Security Issues like Hacking:**-While digital payments support a number of economic projects, it has also led to an increase in cybercrime and financial thievery.A 2017 survey claims that Indian consumers lost over 18 billion US dollars as a result of cybercrime. Over 27 thousand instances of cybercrime were reported in the nation in 2018, a rise of more than 121 percent from only two years prior.
- **Lack of Trust among the Public:** There are certain risks involved in undertaking online transactions like fraud, hacking, identity theft etc. affecting public sentiments and trust in the digital payment mechanisms. There have been instances where even after taking proper precautions, individuals became victims of online payments and have lost all their hard earned money. This prevents merchants, common public etc. from fully relying upon digital payments as a mode of undertaking commercial transactions.
- **Lack of Accountability on and of App Developers/ Intermediaries:** Online payments requires intermediaries who provide a platform for undertaking e payments through Payment Apps. App users are required to create an account linked with their bank accounts for undertaking e payment facilities provided by such apps. It becomes essential to ensure that the app is being protected from any unauthorized access. They must strictly comply with the rules and regulations dealing with online payments, especially know your customer (KYC) requirements such as *Master* Direction on Know Your Customer (KYC), 2016 for ensuring customer identification.

Furthermore, *"Master Direction on Issuance and Operation of Prepaid Payment Instruments" was* issued by RBI in 2009 by virtue of S. 18(Power of Reserve Banks to give directions) read with S 10(2) (Power to Determine Standards) of the Payment and Settlement System Act, 2007 for regulating the framework of Digital Wallets. Furthermore as per S. 3.5 and S. 3.6 Prepaid Payment Instruments (PPIs) – Operational Guidelines for Interoperability dated 16 October, 2018 *"PPI issuers shall adhere to all the requirements of card networks / UPI including membership type and criteria, merchant on-boarding, adherence to various standards, rules and regulations applicable to the specific payment system such as technical requirements, certifications and audit requirements, governance, etc."* and *"a) PPI issuers shall ensure adherence to all guidelines / requirements of card networks / UPI in terms of reconciliation of positions at daily / weekly / monthly or more frequent basis, as the case may be. b) PPI issuers shall adhere to all dispute resolution and customer grievance redressal mechanisms as prescribed by the card networks / UPI."* This makes Digital Finance providers including digital payment app developers liable for their activities.

Suggestions

As mentioned above there are certain limitations/challenges required to be catered for ripping the benefits of Digital Payments in India. Some of these challenges can be removed easily whereas some may require time for their solution. There are few suggestions which can be used in dealing with such challenges and thereby, making Digital Payments in India more secure and efficient. Some of them are as follows:-

- Present legal regime is not too efficient in detecting, curbing and punishing the frauds involved in digital payments. This leads to scams and causes severe financial losses to the individuals thereby, severely affecting the trust of the public on digital payment mechanisms. Thus, the need of the hour is to make the present legal regime regulating digital payments more effective and efficient. Stringent and well defined laws and policies governing digital payments shall be made. The subject matter of such laws should not only be limited to the end users of digital payment mechanisms but shall also include other stakeholders like financial institutions including Fintech companies like Paytm etc. Thereby, making such stakeholders accountable for their business operations.
- Apart from introducing safeguard mechanisms like two-factor authentication, stress must be laid by App developers towards ensuring that payment data is encrypted while transmitting signals between terminals, servers or devices.
- Cyber Security protocols are required to be strengthened for securing digital payments, thereby not only reducing cyber crimes but also helping in gaining trust of the Public on Digital Payments.
- Technological innovations like iris scan, thumb and finger print authentication should be promoted.
- Stakeholders like Government and Financial Institutions should actively try to spread awareness among the general public about the benefits, risks and steps for mitigating such risks through various media platforms. People should be made aware about the success stories and bad incidents that took

place with the individuals while using digital payments. This will not only motivate the public to opt for digital payments but will also make them cautious while making such payments in order to avoid risks.

- Government and financial institutions along with Fintech companies like Bharat Pe shall enter into collaboration with each other with a view to promote digital literacy including financial literacy among the general public. Seminars and training sessions may be conducted with the purpose to train and inform the public about how to make digital payments. Digital media platforms like Youtube, Facebook etc. can be used for promoting literacy on digital payments.
- Concessions and rebates must be given to the retailers, shopkeepers, merchants and other suppliers who use digital payments for their commercial transactions. This motivates merchants to opt for digital payment mechanisms in their business operations.
- Smooth multilingual grievance redressal mechanism must be established so that language does not become a barrier while undertaking digital transactions.
- Multi Stakeholder partnership including customers, financial institutions, law enforcement agencies and central database administrators.
- Strengthening of envisaged CPFR (Central Payment Fraud Registry) database against current regulatory structure comprising of CFR (Central Fraud Registry) for reporting fraud instances of more than 1 lakh corpus.

Conclusion

Digital Payments ever since its introduction in the Indian Economy has been playing a major role in promoting commercial transactions and motivating the public to get connected with the banking system of the nation. Digital payments by bestowing various benefits like ease in recording commercial transactions, efficient and time saving payment mechanism etc. has removed the drawbacks in traditional cash based transactions. Thereby, slowly surpassing cash transactions in the Indian Economy. However, no matter how efficient and sound digital payments appear to be, they still suffer from certain limitations involving riskier payment mechanisms with issues of frauds, lack of data security and lack of stringent digital payment regulating laws etc. which pose to be major challenges for the future of digital payments in India. Such challenges must be eliminated if we want the Indian Economy to be a sound digital economy.

References

Athique, A. (2019). A great leap of faith: The cashless agenda in Digital India. *New Media & Society, 21*(8), 1697-1713.

Goswami, H. (2016). Opportunities and challenges of digital India programme. *International Education and Research Journal, 2*(11), 78-79.

Gurumurthy, A., Chami, N., & Thomas, S. (2016). Unpacking Digital India: A feminist commentary on policy agendas in the digital moment. *Journal of Information Policy, 6*(1), 371-402.

Mukherjee, R. (2019). Jio sparks Disruption 2.0: infrastructural imaginaries and platform ecosystems in 'Digital India'. *Media, Culture & Society, 41*(2), 175-195.

Singh, S., & Rana, R. (2017). Study of consumer perception of digital payment mode. *Journal of Internet Banking and Commerce, 22*(3), 1-14.

❑❑❑

10

The Q-Commerce Upswing

Vandna Zutshi Bhat
Assistant Professor, Manav Rachna University

Dr. Subodh Saluja
Associate Professor, Manav Rachna University

The Pandemic entered the world stealthily in Dec-Jan 2019-2020 and by March it had started knocking on the doors of the citizens of India. The uncertainty, fear of unknown and dismal state of healthcare ensured that the GOI took the tough stance of declaring lockdown from 25th March 2020.With its phases waxing and waning along with the severity and spread of the disease the total lockdown became the harbinger of new ideas. Urban India was suddenly confined to home and the workplace had a new address with office creeping into homes. The time demarcation between personal and official started blurring. With the friendly neighborhood kirana shop closed the attention of an Urban Indian was captured by various ecommerce sites. Soon the only sign of economic activity on the deserted streets and roads of Indian cities under lockdown were the e-commerce delivery persons rushing from home to home to deliver groceries ,vegetables and medicines to consumers locked up in their homes as well as to their computer screens. Even as the lockdown was gradually lifted the habit of ordering online persisted. The birth of a lazy economy during lockdown was here to stay.

The Goliaths of E-commerce players like Amazon, Flipkart, Big Basket and Grofers were already embroiled in a fierce competition. To carve a niche, players like Swiggy instamart, Zepto and Dunzo came up with the unthinkable - 10 to 30 minutes delivery proposal to become the modern day Davids of Q-Commerce. The electronic commerce world has now morphed into a quick commerce world. The genesis of Q commerce has come at a time when the online grocery market in India has been portrayed to grow exponentially and will soon become a $5 billion industry by 2025 that is 10-15 times the present value.

As a country of 1.2 billion people India accounts for the third largest grocery market worldwide.This market is littered by Dad and Pop stores aka the friendly neighborhood kiryana shop which caters to the grocery needs of the customers within its spatial range of influence. The lockdowns during the pandemic proved to be the gate crashing moment for online grocery stores to penetrate these areas which were under the spell of a kiryana shop. The fear of the disease and customers staying away from physical or social interaction of any form was a boon for the online players. With a market of such a large scale there is a high level of opportunity cost in going for online penetration. Thereby ushering in dire need of a differentiating factor that can levitate the percentage of share of this huge market. The fast version of on demand grocery delivery is a classic example of a confluence of the traditional e-commerce and the latest innovation in last mile delivery. Using this premise of fastest delivery as the differentiator were companies like Blinkit (formerly Grofers), Swiggy Instamart, Zepto and Dunzo to name a few. The severity of the pandemic has abated by 2022 but the consumers having tasted blood were reluctant to get over the simplicity and convenience of having unplanned purchases home delivered within 10 to 30 minutes. A consumer craving for a late night ice cream treat or looking to purchase a missed ingredient from the weekly shopping list recollected at 10 pm knew that the proverbial magic lamp of a Q-commerce app would grant the wish. Gradually the mid to high income stratas of the population in big cities have embraced various Q- Commerce platforms while ditching their local Kiryana store trips. The speed of delivery has led to an increasing mid to high income consumer base which is enticing for Q-commerce players. According to the RedSeer study there is a ready pool of 20 million households that is ready to embrace the convenience centric Q-Commerce delivery model amounting to 7% of overall market share. The expected growth rate of 12 to 13% by 2025 makes this an attractive proposition for the existing Q-commerce portals. About 50 % of the exponential growth will be associated with affluent households of metros and Tier-I cities.

Into the foray are the freshly minted Zepto raising about $160 million after two rounds of investments from investors like Nexus Venture partners and Y Combinator thereby increasing its market evaluation to $570 million within a short span of two months. Grofers responded by morphing into Blinkit and Swiggy the food delivery champion set up Swiggy Instamart to test the waters with an investment of $700 .Tata Group owned BigBasket also introduced BBnow to the space. Joining the Bandwagon is Reliance backed Dunzo Daily. Amazon so far has been reluctant to join the race nevertheless is catering to orders at the earliest in two hours via Amazon Fresh.

Speed is the king in the world of Q commerce but how exactly are orders executed in astonishing 10 to 30 minutes. The secret lies in the novel concepts of dark stores and the power of artificial intelligence based predictive supply chain models being followed by various Q- commerce players. The concept of using dark stores is not a very recent phenomena as companies like Walmart and Target in the USA have been using dark stores for curbside picking, in store picking or home delivery. A definition of a dark store is any small brick and mortar store which is closed to an inflow of customers and is just utilized as a center for fulfillment. Walmart and Target have used dark stores to create a convenience factor for the customer who doesn't want to leave the vehicle or even home for accepting the delivery from the store by an employee delivering to the dedicated parking area or allocated pick up area near the front of the store. This mode of operation turned out to be beneficial during the preference to contactless deliveries in the early pandemic years of 2020 -21.Dark stores have proved to be one bright spot for both the retailer as well as the customer in these troubled times.

Advantages of Dark Stores include:

- Quick and contact-free shopping
- Enhanced distribution power and faster delivery
- Considerably large audience and wider scope
- Better SKU management
- Trouble free perishable products management
- Capacity to stock a wider range of products
- Upgraded inventory control and management

With all these favorable aspects the Q commerce companies were quick to employ the strengths of Dark stores to consolidate the fast delivery mode of business. The dark stores are strategic nodes in the development of hyper local supply and delivery networks within a 2 to 2.5 km range. There is absolutely no way that fast deliveries in minutes are possible in a dense urban sprawl otherwise. The hyper local clusters of dark stores and the tie ups with the local grocery stores are the spine of the q commerce model without which the speed bumps of heavy traffic and congestion on any large city road would ensure that this model never thrives. Speed is not the only factor governing the performance of the Q commerce model. There are hidden attributes like an efficient inventory management system along with pick accuracy, predictive shopping list as per last purchase and timely handover to the delivery person which levitate the operations to overcome any retarding friction which could transpire during an order execution.

A tour of a typical dark store operating in a dense metropolitan neighborhood will reveal a watertight model of a mini warehouse assisted with smart devices enabled by artificial intelligence. It is a quiet place where all the talking is done in digital mode. The short pick time to collect the items from the shelves is again due to intelligent Inventory management in place. The inventory maintained is of those items which belong to a typical urban home grocery list. This includes the perishable items also as these items do not have to stay long enough in the stores to risk spoilage. Some Q commerce companies also use predictive analysis to forecast the expected order based on the time of the day, location of the store and the week day. Data from such predictive analysis throws up interesting facts such as the tendency to order items on a weekend while commuting from office to home. The data also shows a spike in the number of orders received post lunch and in the evenings The AI enables the system to identify frequently ordered items for each of its regular customers and prepare a predictive list of items for storage and picking. The efficient SKU management plays a very important role in speeding up the process of picking, packing and dispatch. As the store is closed to the inflow of customers the space inside can be utilized to incorporate a denser layout of shelves with very narrow aisles thereby augmenting the capacity to store a heterogeneous range of products. The products are stacked neatly on the shelves .Emphasis is on the prober barcode labelling of shelves containing different products rather than on the aesthetics or display as would have been the case in a typical supermarket a typical dark store can cater to a 5 km radius. The location can be any non-descript part of the urban scape. The exterior rarely reveals the orderliness and the fluid moments taking place inside by the ever busy pickers. The pickers are mostly young and in their 20s. A basic 12th standard education is enough for hands -on work, though many graduates are also lured by this new economic movement. The pickers work in shifts to keep the action going. Even the lunch time is coordinated in a one at a time manner to maintain the tempo. A typical picker can earn anywhere between Rs 18000 to Rs 20000.Some Q commerce companies also have the incentive based payment model where speed picking, packing and dispatch are rewarded. The constant humming of the air conditioners meant to keep the produce fresh ironically improves the working conditions for the employees zipping around the store to collect and pack each order within the stipulated 1 to 2 minutes.

When an order lands from the Mobile app or from the e-commerce website onto the computer, it is picked up by a smart hand held device. The deceive gives out a shrill beep. As soon as the picker picks it up, like a music conductor it guides the movement of the picker through various narrow aisles to the required items while scanning the barcode tags on the shelves. These items are quickly removed and placed in the collection tray by the picker. The device then communicates with the computer for billing. While the billing process is underway two activities are taking place simultaneously: packing and rider movement towards the store front. Packing is the only time where pickers put their vocal chords to use while double checking the items as per the customer order. Packing can take approximately 20 to 30 seconds and that is the time period within which the rider gets the notification and arrives to pick up the order. Soon the order is on its way to be delivered to the eagerly awaiting customer within the promised 10 to 30 minutes subject to the recalcitrant traffic conditions. The picking, collecting and packing has to be completed within the 1 to 2 minutes window for the order to be executed within 10 to 30 minutes.

The promise of speedy delivery does not ensure market dominance, the competition is very stiff, and within a few days of a dark store opening in an area the competitor also sets up shop. This makes the 10 to 30 min delivery promise very vital to retain the customer base. The technology and human efficiency does make this possible to achieve but there are quite a few speed bumps to be encountered. Every big city where express delivery is the new customer infatuation with convenience, it is the Traffic management department which bears the brunt of delivery rides zipping through red lights, driving rashly to meet the speedy delivery targets. Delivery riders are proving to be a forebears of hazardous traffic behavior in all big cities already suffering from choking roads and insufficient infrastructure. This has led to a few voices being raised against the model. Some are tagging this as a speed driven incentivized form of exploitation of all the employees working for a Q commerce company. Recently Trinamool Congress MP Mohua Maitra sought a ban on such 10min delivery gigs which can endanger lives of riders as well as other commuters using the roads. During the initial phase riders get around Rs25 to Rs 35 per ride and with the experience the commission is raised to Rs 50 per ride. The target of each rider is 40 rides per day which means earning of about Rs 1500 per day. For many of these riders this has been a new lease of life to run their households as most are those who lost their jobs to the vagaries of the pandemic. Thus the incentive to take as many trips as possible and the pressure to deliver within 10 minutes becomes a back breaking work which is difficult to sustain in the long run.

The demand for the Quick Commerce is expected to expand with at a double digit of CAGR on the basis of GMV between FY'22P and FY'27F.Technological innovations have already made inroads into the supply chain and distribution networks all over the globe E-commerce warehouses are replete with robots in picking area and AI based warehouse management softwares working for seamless operation. Drone technology being used to deliver packages will be a game changer. Test runs using drones are already being undertaken in some cities in India. To run a future dark store automation will be integral along with drone Technology, voice ordering, and use of Electric Vehicles for delivery will drive the Quick Commerce Market in future. Demand Prediction and Customer Behavior Analysis is crucial to ensure sufficient availability of the groceries and to ensure minimum degradation and stock-outs of these products. Quick commerce as a business model is here to stay and is poised to morph from a lazy model of convenience shopping to the new normal of an urban lifestyle.

SWOT Analysis

STRENGTHS	**OPPORTUNITY**
Quick convenient and contact-free shopping mode for staples Enhanced distribution power and faster delivery Trouble free perishable products management Upgraded inventory control and management Cost Effective No time constraints -24 hours*7 days a week delivery	Caters to large segment of customers over concentrated geographical area in Tier 1 and tier 2 cities Considerably large audience and wider scope Better SKU management Increasing number of users gaining trust in this delivery model Online payments mechanisms like Paytm speed up the order placement process High availability -24 hours*7 days a week Integration of technological innovations

WEAKNESS	THREATS
Limited set of items that can be kept in dark store as compared to a supermarket Frauds - Data breaches can lead to data theft with personal and financial details of customers misused by hackers for fraudulent purposes Fake websites - That promise better service may dupe customers and fly by night operator Touch ,feel and see aspect missing from the shopping experience Limited exposure - In case of internet services issue / not available this services cannot be implemented Speed delivery promise subject to external factors like traffic, mode of transport, speed of pickers Pickers and riders can be exploited	Local kirana stores having their own home delivery model No direct interaction of customer with the seller Too many players in the segment leading to dilution in market share

References

https://www.indianretailer.com/article/technology/digital-trends/how-q-commerce-is-becoming-one-of-the-fastest-growing-e-commerce-models.a7778/

https://www.moneycontrol.com/news/business/visit-to-a-dark-store-heres-what-it-takes-to-deliver-that-bag-of-groceries-to-you-in-10-minutes-7849131.html

Sunday Times of India June 19 2022 edition

❑❑❑

SECTION - 4
Banking & Finance Sector

11

Digital Banking and Fintech

Ramita Yadav
Assistant Professor,Department of Commerce, Maitreyi College, University of Delhi

Dr. Ashima Saxena
Assistant Professor, Senior Scale, The NorthCap University

Abstract

This chapter focuses on the area of digital banking as well as on fintech. Services provided by digital banking are of great importance in the modern era. These services help customers gain better experiences. This chapter throws light on Digital Banking Units (DBU) as the Government is planning to set up 75 DBUs in near future. Services that will be offered by these DBU's are explained briefly. Financial industry has changed a lot in the last decade. Technology helped in evolving this area. And a new industry has been emerged ie. Fintech industry. Evolution of the fintech industry is stated in this chapter. Different areas of fintechs are explained such as peer to peer lending, cryptocurrency, blockchain technology, crowdfunding, artificial intelligence, underwriting etc. Regulatory framework for this newly emerged area is a challenge for regulatory bodies also. In this context RegTech, SupTech and Regulatory Sandbox (RS) framework is discussed. And finally a way forward for the traditional banks and financial institutions is emphasized.

Keywords: *Digital Banking, Fintech, Peer to Peer lending, Artificial Intelligence, Blockchain*

Digital Banking

Meaning

Traditional banking provides financial services only with the physical presence of customers. It consumes both time and cost.

Traditional banking has a lot of paperwork and risk of carrying cash by customers is also present. For every financial service customer has to visit the bank. However to provide customers a better experience financial activities are taking a digital form. Banks provide some of the financial products and services through digital mode also. It is called Digital Banking. Banks are facing tough competition due to the financial innovations emerging in the financial market. Financial innovations are very attractive to the customers and banks can lag behind if banks do not adapt to these changes. Therefore, Digital Banking is a way out for banks to accept a new financial environment.

Services Provided in Digital Banking

1) Fund Transfer

Online fund transfers are possible by using digital banking. There are different modes available for fund transfer which are as follows :

- **NEFT (National Electronic Fund Transfer):** Banks offer customers to send money through NEFT online mode also. However, while using NEFT beneficiary customers get money on half-hour batches and immediate payment can not be done by using NEFT online mode. There is no minimum and maximum fund limit available for NEFT transactions. Service is available for 365 days 24*7. If NEFT is done using online mode then transfer charges are not levied.
- **RTGS (Real Time Gross Settlement):** Banks provide the facility to transfer funds through RTGS online mode also. RTGS allows beneficiary customers to get money immediately and service is available for 365 days 24*7. If RTGS is done through online mode then no service charges are levied however minimum amount to be transferred should be 2 lakh or more than 2 lakhs for using RTGS. RTGS is now open for Fintech players also such as for Paytm mobikwik phonepe etc.
- **IMPS (Immediate Payment Service):** IMPS can be availed through online mode only. Bank charges some fee to transfer the money immediately. Maximum amount that can be transferred using IMPS is Rs. 2,00,000. Service is available 365 days 24*7.
- **UPI (Unified Payments Interface):** The NPCI (National Payments Corporation of India) launched UPI in 2016. It operates on IMPS (Immediate Payment Service) and is basically an API (Application Programming Interface). It transfers

funds immediately in the real time. Inter-bank person-to-merchant (P2M) and peer-to-peer (P2P) operations are fostered by the UPI. UPI integrates multiple banks into an only one smartphone application. RBI (Reserve Bank of India) governs Unified Payments Interface. Popular UPI's : Paytm, Google Pay, PhonePe, Mobikwik etc.

- **BHIM (Bharat Interface for Money):** BHIM was launched by NPCI (National Payments Corporation of India) in 2016. It is a payment mobile application. BHIM is an Indian app. Government created this app to achieve the goal of financial inclusion. It directly sends the money from one bank account to another bank account. Point should be noted here that unlike paytm or other UPI's it does not have any wallet. BHIM can transfer money without internet connection also by dialing *99# and thereby rural area population can also benefit from this fund transfer method.
- **AePS (Aadhaar-enabled Payment System):** It is a service that allows account holders to utilize the aadhaar number to operate basic financial activities. For example, checking bank balance, deposit or withdrawal of money. AePS is done through a business correspondent. A business correspondent is an approved bank agent. Business correspondent uses a micro ATM (a hand-held device) which by using aadhar number, fingerprints of the customer traces the account of the customer and to send money it maps the beneficiary account through beneficiary ID. There would be no fraud committed since it relied on biometric data authentication. It is basically electronic fund transfer from aadhar to aadhar.

2) Net Banking

Net banking is also called internet banking. Different banks provide different services through internet banking. Account opening or closing can be done, insurance schemes can be purchased, PPF accounts can be linked online through internet banking. RTGS, NEFT and IMPS can be done by adding beneficiaries in netbanking. Detailed account statements can also be viewed using netbanking.

3) Mobile Banking

Mobile banking is used with the help of applications in smartphones. Many banks have their apps particularly made for customers to enable them to use mobile banking. Customers can check their balances, transaction history etc. through mobile banking apps. It is very easy and convenient for bank customers to use.

4) Demat Account (Dematerialised Account)

Demat account helps customers to have their securities holdings in electronic form. All investments are recorded at one place on an online platform. Shares are traded in electronic form instead of physical certificates. Depository participants help in converting physical shares into Demat holdings. As shares are available online, it is easier to access them from anywhere for customers.

5) Other Services

Digital banking offers customers to obtain bank statements virtually, cash withdrawals by ATM's, payments of bills using digital platforms, managing personal finance digitally and can monitor transaction records online.

Benefits of Digital Banking

- It provides ease of banking from the solace of one's own home.
- It reduces the demand for fiat currency.
- Digital banking allows banking operations to be available 24 hours a day, seven days a week.
- Digital banking allows to set up direct debits for the regular bill payments.
- It restrains the movement of black money.
- It is banking without the use of paper.
- It can render services to remote locations also.

Limitations of Digital Banking

- **Cyber crime:** Digital banking is prone to cyber crime. Cyber crimes are very hazards for individuals using banking via online platforms. Cyber crime activities include trying to obtain users' ID, passwords, private financial information or one time passwords etc. to reap financial advantage. System or mobile phone applications also get hacked to obtain private monetary information. Cyber crime is very threatening as in the case of cyber crime proofs are generally harder to trace as well as these are complex crimes which a layman can not understand easily.
- **Internet connection:** There exists a hidden cost while using digital banking ie. price paid for internet connection. Internet connection speed varies at different points of time. Slow speed of the internet can affect financial transactions. On top of internet connections, people need to depend on the system's performance and reliability too. Users can not log in to bank

accounts if the internet connection is unavailable for an extended period of time due to technical issues. Users can not use digital banking if their bank's site maintenance is going on.

- **Human relation is missing:** In the digital banking human interface is missing or it is very complex to get in touch with customer service providers. And even if customer service is available, service providers do not know the user's banking history. In addition to this, customers frequently have to navigate a phone tree and wait on hold before speaking with customer service providers.
- **All services can't be availed:** All banking services can not be availed using digital banking. Customers have to visit bank branches for various services such as opening of account, applying for a loan, purchasing products from a point where only cash is accepted etc.

Digital Banking Units (DBU)

A digital banking unit is a customized fixed point business hub that will have a minimum technology platform for providing digital financial services and products and also providing conventional financial products and services digitally in self-service mechanisms at any time.

Scheduled commercial banks will open 75 digital banking units in 75 districts of India as proposed by the budget of financial year 2022-2023 to celebrate 75 years of the independence.

Set up

Commercial banks which are having prior experience of digital banking can open digital banking units without seeking the permission from RBI.

Digital Banking Units Services

- It must provide digital kit to customers as well as to merchants.
- DBU will provide UPI QR code and BHIM Aadhaar service to the customers.
- DBU's will provide online saving bank accounts under various schemes, online RD(Recurring deposit) accounts, online FD (Fixed deposit) accounts and online current accounts.

- Customers will be offered mobile banking, internet banking.
- Debit cards, credit cards and mass transit system cards will be offered to users by DBU's.
- Point of sale (PoS) service can also be availed.

Fintech

Meaning

Fintech (Financial technology) are the institutions which provide financial services by using technology. Artificial intelligence and machine learning technology made it easier for fintech companies to provide financial services in various fields like education sector, banking services, investment sector etc. Fintech is the combination of financial service providers with information technology companies.

Evolution

To understand the evolution of the Fintech industry globally, we need to study the following phases :

Phase 1 : Early 90's to 1970

Financial services were offered through traditional mode in the early 90's. However, the development of computerisation globally developed the fintech industry also. Analogue technology is the identity of this phase.

Phase 2 : 1970 to 2008

Usage of the internet leads to digitalisation of financial products and services. Many traditional financial entities offered services through online mode during this period. Launching of ATM's in the market is a major financial innovation in this period. ATM's made withdrawal of money very convenient. Technology shifted from analogue to digital. But the 2008 global financial crisis changed the scenario a lot. Regulatory reforms after the global financial crisis shaped the fintech industry.

Phase 3 : 2008 onwards

There are approximately 2100 fintech companies presently. Out of which almost 70% of the companies have grown in the last 5 years. Many startups have come in the market in this phase. Artificial intelligence changed the scenario of the fintech industry substantially.

Fintech Areas

1) Peer to Peer Lending

Peer-to-peer lending is the process of loaning money straightforwardly to individuals or organizations without the involvement of a formal banking institution as a mediator. Fintechs connect borrowers with lenders and provide services like Identity management, loan maintenance, loan approval etc. Individuals and corporations can borrow and lend money to one another through peer-to-peer lending.

2) Cloud computing

Cloud computing is a set of digital infrastructure such as data centers, flash memory, records, connections, operating systems, data science, and insight via the Web in order to supply faster innovation, quite versatile funds, and scale savings. Users generally pay only for the cloud services users use, limiting their expenditures, enabling you to operate their infrastructure quite productively, and letting users to expand as one's market conditions change.Banks are able to collaborate with fintech firms using cloud computing, and innovators have been working to develop as cloud native from the beginning.

One of the benefits of cloud computing for fintech companies is data aggregation. It is vital to safely store a certain findata such as bank balance details, purchasing patterns, financial planning, and money flow. Collecting information from financial sector datasets allows for proper refining. This findata's ease of access and privacy are immensely helpful not simply for banking firms but to customers also.

3) Machine learning

Fintechs can use machine learning algorithms to anticipate market volatility, identify expected investment opportunities, and prevent fraud etc.. Companies can train machine learning models on enormous amounts of information like monetary transactions, stock portfolio, mortgage repayments, customer interactions, and so on. It thus ensures that they are able anticipate emerging trends in financing, equities, and so on.

4) Artificial intelligence

The incredible spread of artificial intelligence throughout businesses is possessing a remarkable effect on the socioeconomic, and workforce contexts. Fintech contenders involve robo advisory services powered by AI (artificial intelligence). Robo-advisors are web - based platforms of participatory and perceptive user assistance elements which use digital technologies to help customers through an automated trading consulting procedure. The expansion of AI-based fintech companies has prompted numerous acquisitions and mergers.

5) Crowdfunding

Crowdfunding is the use of small sums of money from a large number of people to fund a new business venture.

Types of crowdfunding business models by fintechs:

1) Charitable model

Non-profit entities use the Charitable model to secure funding for generous or divine causes. Funders are faithful, and founders will recurrently give them information of the charitable venture's growth in order to make repetitive charitable contributions.

2) Online lending models are platforms that permit borrowers to acquire borrowings and investors to buy note cards supported by loan payments. The platform serves as a go-between. Compliance and governmental law applies to online lending.

3) Pre-sales crowdfunding model

A service/product is placed online in a Pre-Sales crowdfunding model, and funding sources are inquired if they would like to place an order for these in advance.

4) Equity crowdsourcing model

When a producer seeks to raise seed funding from the crowd rather than financial institutions, an equity crowdsourcing model can help.

5) A Rewards-Based Model:

A Rewards-Based Model is ideal for entrepreneurs who are seeking to raise funds for a small business by offering non-monetary incentives.

Popular fintechs in crowdfunding are:

- **Indiegogo:** This popular crowdfunding website initially raises money for independent films only, however it now raises money for different categories like : Health & fitness, phones & accessories, travel & outdoors, audio etc.
- **Kickstarter:** Kickstarter is the crowdfunding website for startups looking to raise capital. In fact, Kickstarter can only be used to create projects that can be shared with others. And it does not allow startups to use charitable mode or equity crowdsourcing mode.
- **GoFundMe:** GoFundMe is also a popular fundraising website for people who need help recovering from a medical expense

or disaster, such as a natural disaster, house fire,or any other emergency expense.

6) Blockchain

Blockchain is defined as a decentralized, distributed public ledger or it is a public database of digital information. Blockchain is a particularly secure method of data storage because it is nearly impossible to manipulate without being observed.

Anyone on the globe with access to the internet and a smartphone can just use DeFi firms, DeFi (Decentralized finance) is literally the union of blockchain with fintech, to create a global, open alternative for all of the established everyday financial instruments, like trading, savings and loans, and insurance. This is the present situation, which bypasses the need for a middleman and allows loans to strangers, and more extensive trading techniques.

Blockchain aids FinTech by providing a new dimension of freedom and security through the distributed, decentralized, transparent, and unchangeable digital ledger technology. Information does not fall under the control of a single body. The data cannot be altered by anybody. In addition to being robust, it promotes actual equality and democracy. This is a fantastic opportunity to offer banking services to billions of people all over the world who would otherwise need to go to conventional banking or unauthorized markets.

The speedy issuance of digital securities at lower unit costs and with more customization is facilitated by blockchain-based inclusive, open, and secure business networks. In the financial sector, blockchain technology has evolved and shown benefits including transparency, security, trust, privacy, programmability, high performance, and scalability.

7) Cryptocurrency

Cryptocurrencies are digital currencies which are not regularized by any Central Government authority and that is why these have a decentralized nature. As these are decentralized currencies, its verification is done by blockchain technology. Bitcoin is also a popular cryptocurrency.

Traditional banking services are rendered at a slow pace as there are multiple levels to go through before the completion of transaction, however cryptocurrencies can be transferred very swiftly as these are decentralized and can move back and forth very quickly. Speed, visibility, and comfort are the pillars of Fintechs, and cryptocurrency is a component of this. As blockchain technology is reliable, all such data cannot be exploited, making fraud prevention a less expensive and challenging entity for Fintechs.

8) Underwriting

In the earlier times, the process of underwriting was not digitized and not equipped with technological solutions and hence took a lot of time to evaluate customers' credit rating. The credit rating was used as a preliminary step to underwrite or evaluate a person's financial health or ability to repay a loan. It becomes very difficult to assess a customer if they have a thin credit history. Fintechs have entered the lending environment, constantly evolving its trajectories over the last couple of years. Fintechs employ alternative credit rating pathways to analyze a borrower's online activities in order to ascertain financial health. Data is collected from telephone, insurance, electricity ,mutual fund companies etc. and thus have a broader range of data than traditional financial institutions.

Social media is also used to gather data for customers as it can tell about their behavior, likes, hobbies,background etc. and helps in assessing the intention of the borrower in repaying the loan amount.

9) TREDS (Trade Receivables Electronic Discounting System)

TREDS is a new technology developed for discounting bills of MSME (Micro,Small and Medium Enterprises) using electronic platforms. TREDS helps MSMEs to convert their receivables into liquid funds. Currently there are 3 TREDS platforms namely : Mynd Solutions or M1 Exchange, A. TREDS Ltd. or Invoicemart, Receivables Exchange of India (RXIL). By discounting bills on TREDS MSME receives money on a prompt basis and MSME need not to wait for the due date to arrive of invoices. Parties involved in TReDS are buyers (Companies, Public sector undertakings and bodies as permitted by RBI), funders (Banks, NBFCs, Fintechs, and bodies as permitted by RBI) and vendors (MSME) as per MSMED Act, 2006.

Traditional lenders discounts bills of medium and small enterprises which are almost 6% of the MSME sector. Whereas a large portion of 95% micro enterprises in the sector MSMEs remained deprived of the facility. However, the government after facing pandemic has allowed some high value fintechs to be a part of this platform so as to meet MSMEs priority sector lending liability. These fintechs are gaining attention due to their data-driven models and do not require collateral securities, documentary evidence etc. which micro enterprises do not possess generally and hence helping micro enterprises to get the benefits of TReDS platform. However, till now approx. 150 fintechs are allowed to enroll on the platform. Government should allow more fintechs on the platform and help micro enterprises to meet their funding requirements.

Regulatory Framework for Fintechs

RegTech

RegTech or Regulatory Technology helps fintechs to comply with regulatory norms. It also equips regulators to understand new technologies and form the regulatory norms accordingly. There are a number of RegTech solutions available in different sizes.

SupTech

SupTech is a type of Regtech which enables regulators to verify and supervise regulated institutions. SupTech checks for the regulatory norms adopted by fintechs.

Regulatory sandbox/Innovation hub

RS (Regulatory sandbox) Framework is used by RBI to check new financial products provided by fintechs in the market. Fintechs are selected to test their products in the regulatory sandbox if they qualify certain eligible criteria such as proper IT infrastructure, net worth of at least Rs.10,00,000 etc. RS framework only tests products that are new or existing products with modification.

Way Forward for Traditional Banks

The EASE Reforms Agenda was introduced collectively by the govt and Public sector banks in January 2018. It aims to institutionalize smart banking. EASE reforms will be introduced in 4 levels. EASE reforms majorly focus on digitalisation, technology adoption by banks, partnerships with fintechs and digital agriculture financing.

EASE 1.0:

Report found a significant advancement in the transparent settlement of Non-Performing Assets.

EASE 2.0:

It expands the above reform by introducing new reform action points for six areas namely :

- Responsible Banking
- UdyamiMitra
- Digitalisation
- Governance
- User Responsiveness
- Credit Off-take

EASE 3.0:

It aims to improve the ease of banking in user experiences through the use of technology, such as partnerships with Fintechs, Credit at click, tech-enabled agriculture lending etc.

EASE 4.0:

Banks will concentrate on co-lending with non-banking entities, digital agriculture financing and technological adoption for 24x7 banking.

The only way forward is digital. Banks will have to jump on board by implementing cutting-edge technologies such as Deep learning, Machine learning, Artificial intelligence, Virtual currency, Internet of Things and the Blockchain.

References

Anil, K., & Misra, A. (2022). Artificial intelligence in Peer-to-peer lending in India: a cross-case analysis. *International Journal of Emerging Markets.*

Abid, S. (2016). Electronic Payment System: An Evolution in Indian Banking System. *IOSR Journal of Economics and Finance*, *7*(2), 25-30.

Chatnani, N. N. (2018). Receivables management and supply chain finance for MSMEs: Analysis of TREDS. *Academy of Strategic Management Journal*, *17*(3), 1-8.

Cook, W., & Raman, A. (2019). National Payments Corporation of India and the remaking of payments in India. *Consultative Group to Assist the Poor Working Paper.*

Das, S. (2019). Opportunities and challenges of FinTech. *Keynote Address Delivered at NITI Aayog's Fin-Tech Conclave, New Delhi*, *25*(3), 2019.

Gupta, A., & Arya, P. K. (2019). Banking Reforms in India with Special Reference to Digital Banking. *Asian Journal of Research in Banking and Finance*, *9*(7), 12-27.

Haralayya, B. (2021). How Digital Banking has brought innovative products and services to India. *Journal of Advanced Research in Quality Control and Management*, *6*(1), 16-18.

Joshi, M. (2017). Digital Payment System: A Feat Forward of India. *Research Dimension (ISSN: 2249-3867).*

Manda, V. K., & Yamijala, S. P. (2019). Peer-to-peer lending using blockchain. *International Journal Of Advance Research And Innovative Ideas In Education*, *6*, 61-66.

Ma, Y., & Liu, D. (2017). Introduction to the special issue on Crowdfunding and FinTech. *Financial Innovation*, *3*(1), 1-4.

Martínez-Climent, C., Zorio-Grima, A., & Ribeiro-Soriano, D. (2018). Financial return crowdfunding: literature review and bibliometric analysis. *International Entrepreneurship and Management Journal*, *14*(3), 527-553.

Mei, L. (2022). *Fintech Fundamentals: Big Data/Cloud Computing/Digital Economy.* Stylus Publishing, LLC.

Popescu, A. D. (2022). Understanding FinTech and Decentralized Finance (DeFi) for Financial Inclusion. In *FinTech Development for Financial Inclusiveness* (pp. 1-13). IGI Global.

Rao, S. P., & Anand, M. R. (2019). Peer to Peer Lending Platforms in India: Regulations and Response. *Prajnan*, *48*(2), 107-122.

Zetzsche, D. A., Arner, D. W., & Buckley, R. P. (2020). Decentralized finance. *Journal of Financial Regulation*, *6*(2), 172-203.

❑❑❑

12

Digital Finance in India: Growth and Challenges

Dr. Sachita Yadav

Assistant Professor, Arun Jaitley National Institute of Financial Management
An Autonomous Institute of Ministry of Finance

Abstract

When money is in electronic form and transactions like payment and receipts happen electronically, when you need not stand in front of the cash counter in the bank and ATM, when it is taking a few minutes to transfer money from one account to another account it is called digital finance. The current chapter provides an in-depth insight about the three important aspects of digital finance i.e., growth of digital finance in India, challenges in digital finance in India and scope of digital finance in India. This chapter has given an insight about the tremendous development of digital finance in India in the form of IoT, Robotic Data Procession, Digital Payment, cash-less transactions etc. The chapter also discussed the determinants which are hindrance in the way of digital finance like income level, education level, gender gap, rural and older population, unavailability of appropriate digital infrastructure, safety and privacy concern etc. are the bottlenecks in the growth of digital finance in India. The chapter has given ways to outcome from these difficulties and suggested that it is necessary to have an efficient communication network with good connectivity and electricity, focus on literacy and financial literacy, campaign digital finance awareness programmes in rural and semi-urban areas and to build trust among the population about digital finance.

Key words: *Digital Finance, digital infrastructure, security, digital finance awareness programmes*

Introduction

A speedy digitalization in India has changed the face of the Indian economy. In the year 2014, Digitalization, a flagship programme was started by the Indian government with the vision to convert India into a digitally empowered economy and knowledge society. Various initiatives have been taken by the Indian government like Aadhar Generation, Digi Locker, My Gov Portal, e-Hospital Transactions, Jeevan Pramman Pension, Financial Inclusion and many more to convert Digitalization mission into reality. These initiatives have an economic, social and environmental impact for the development of an economy. Digitalization will have a positive impact on the economy by attracting other economies for the investment purpose, enhancement in the output of all the sectors and in accelerating the GDP. The whole society will be impacted positively due to digitalization as the standard of living will improve due to easy access of resources. Digitalization is not only good for the economy and society but also plays a significant role for the environment as well. As it is ng to reduce paper consumption, fuel consumption and travel cost. Figure 1 depicts that India is the second-fastest digital adopter country among major digital economies in the world.

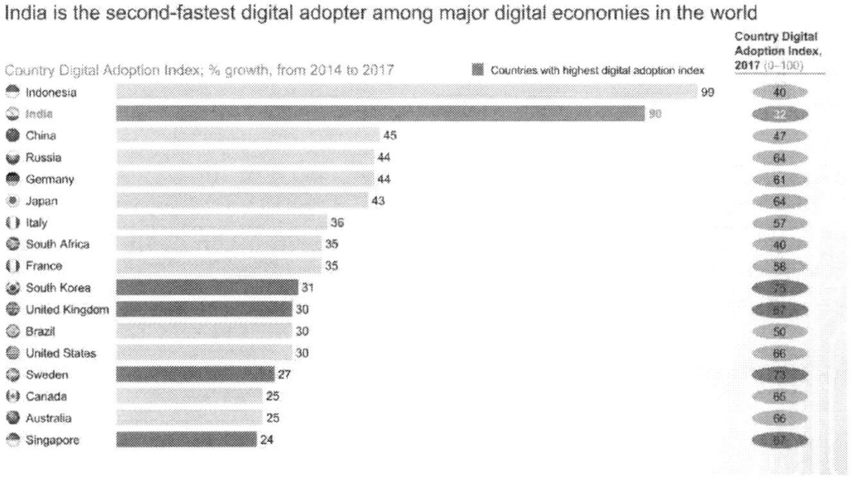

Source: https://pib.gov.in/pressreleaseiframepage.aspx?prid=1565669

Figure 1: India is the second-fastest digital among major digital economies in the world

Financial sector is among the various fastest digital growing sectors in India. Lots of innovation has happened in the financial sector like development of new products, improvement in services and processes, advances in technology, and payment systems and risk management in a greater speed due to digitalization. In the year 2017 a report has been published by World Bank and it has been mentioned in the report that *"total volume of digital transactions in India (including various payment channels and mechanisms, such as net banking, mobile banking, debit cards, credit cards, prepaid instruments, mobile wallets, among others) grew by compound annual growth rate (CAGR) of 30% from 1,142 million in April 2015 to 1,928 million in April 2017".*

The current chapter proposes to study digital finance in India and to understand growth, prosperity and challenges in adopting and implementing digital finance in India. There is a need to build a platform for policy makers and technical analysts to formulate better financial products and services that facilitate the enhancement of digital finance in India.

Finance is the blood of any economy and a good & sustainable financial system is essential for smooth conduct of financial activities in the country at every level (industry level, individual level and country level). There are two important words i.e., financial inclusion and digital finance is playing a significant role for the development and enhancement of financial transactions in the country. To implement financial inclusion and digital finance in any country there is a need to have sufficient infrastructure, legal frameworks and social awareness and attitude and readiness to adapt the new technology. Being a young, dynamic and developing country India is having all the potential to strengthen the financial system by using new technology and has already taken many initiatives like PMJDY, Demonetization, UPI linked Payment System etc. regarding the same. *"India's digital journey is one of exuberance. The country had the world's second-largest internet population at over 749 million users in 2020. Of these, 744 million users accessed the internet via their mobile phones. Estimates suggest that this figure would reach over 1.5 billion by 2040." Tanushree Basuroy, Mar 15, 2022.*

So, there is a wide scope of digital finance in India, as the main features of using digital finance include the capability to use mobile devices, quickly authenticate identity, and real-time payment services. *"Their scalability means that they can be applied to hundreds of millions of customers, even for low-value transactions. Such low-cost, high-volume, low-value digital financial transactions were unimaginable just a decade ago. Much has already been achieved: in India alone, nearly half a billion adults opened bank accounts between 2011 and 2017."* Derryl D'Silva, Zuzana Filková, Frank Packer and Siddharth Tiwari (2019) mentioned that innovation in financial technology like Fintech has given new experience to all of us whether it's a customer or service provider. It helps in providing the better services to the customer in an efficient manner and at lower price points.

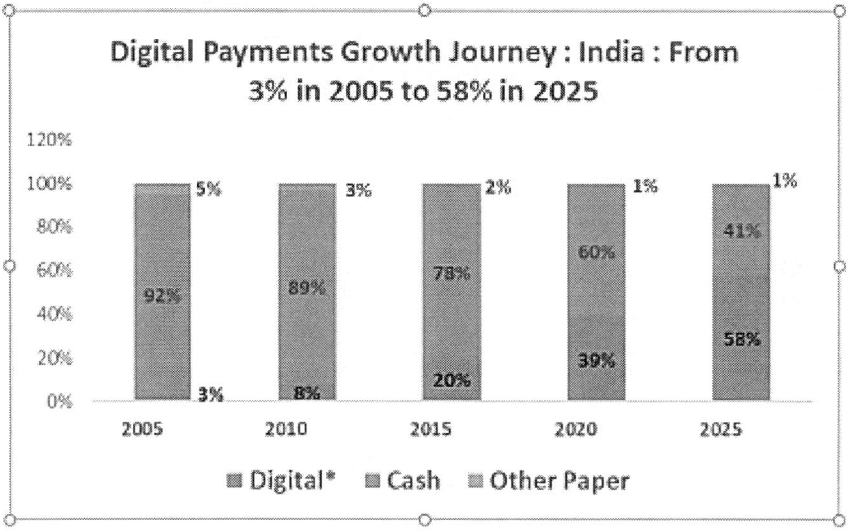

Source: https://community.nasscom.in/communities/digital-transformation/fintech/india-digital-payments-2020-launching-the-first-adoption-index-time-is-now.html

Figure 2: Digital Payment Journey in India

Figure 2 is showing the growth of acceleration in the digital payment system in India. During the year 2005 the there was only 3% of the total payment were through digital mode but surprisingly it has been estimated to increase to 58% in the year 2025, this is showing the popularity of digital payment system in India. The current chapter provides an in-depth inside about the three important aspects of digital finance as mentioned below:

- Growth of Digital Finance in India
- Challenges in Digital Finance in India
- Digital Finance in India- Scope

Growth of Digital Finance in India

The area of finance is very wide. Almost every day customised financial products are being developed in India to meet the requirement of the customers. In the world of digital finance there is no need to visit the financial institution in case of any issue. Facilities like chatbot virtual assistance will be given to the customer in a more effective manner. The growth of digital finance is possible due to the improvement in the infrastructure. Nowadays almost every Indian is having mobile devices with good network connectivity and the enhancement in the system is more efficient. Nowadays predictive analysis is to be conducted with the help of AI to provide real-time solutions to the customer. This is like providing personal services to the customers which was not possible in the past. Nowadays the traditional ways of checking, savings, and loan management processes have shifted to an online system. This provides customized services to individual customers and provides satisfaction to the customer, which helps in improving the customer retention. Automated systems help in checking and keeping the records in electronic form, this is going to help in loan approvals and preventing financial frauds. Neo banking is a new name in the field of banking and this is to create exclusively digital only services where there is no need for physical bank branches. Everything needs to be done electronically and in an automated system. Jupiter, InstantPay, Finin, Zerodha for online trading are few examples of contact less and zero physical presence financial services. Adoption of cloud technology will open a new and unlimited sky for the financial sector. This delivers unlimited and faster ways to provide services to the consumers with less capital investment. Internet of things devices made it easy to analyze and provide customized services to the customers and help in providing accurate information. Augmented reality, quantum computing, peer to peer (P2P) transaction technology and robotic process automation has improved and enhanced the speed for financial transactions. Still the research and innovation are going in these areas for the further improvement and confidently consumers will get the improved financial products in the forthcoming. Ministry of Electronics & IT has reported (Posted On: 20 FEB 2019 7:59PM by PIB Mumbai) that *"India can create up to $1 trillion of economic value from the digital economy in 2025, India is the second-fastest digitizing economy amongst 17 leading economies of the world, Thirty digital themes can be scaled up nationally to accelerate progress in nine priority areas, India's digital economy of the future could generate productivity and output sufficient to support 55 million to 60 million workers in 2025"*, which indicates the growth and prosperity of digitalization in India by 2025. Digitalization is the path for the further development of digital finance in India.

Challenges in Digital Finance in India

Cash is having universal acceptance and the features of cash like liquidity, flexibility, concealment etc. have made the Indian population restrict to the digital finance system at the initial stage. The practice of digital finance is not free from challenges. Till the year 2016 Indian population was not much aware about digital payment and digital financial system. The infrastructure was also not suitable for digital financial services. Chandralekha Ghosh & Rimita Hom Chaudhury (Dec. 2020), mentioned that numerous efforts like demonetization have been made to promote digital finance in India, since 2014. To create a cashless economy financial inclusion has been rolled out in India. The study examined the determinants and impact of demonetization on digital finance. It also examined the gender gap in using digital finance pre and post digitalization and found that men are more educated and technology lovers. Two important determinants i.e., income and education playing a significant role in digital finance and found that women are lagging behind in using digital finance in comparison to men.

Rajesh Barik, Pritee Sharma (2019), analyzed the improvement and forecasts of financial inclusion in India, found that after PMJDY (August 2014) there has been a successful progress in opening bank deposit by the underprivileged people of the society and enhancement in digital transaction after demonetization (November 2016). But digital finance is still beyond the reach of women, rural population, older population as they are less educated and deprived from the technology. Lack of financial literacy, unavailability of appropriate digital infrastructure is also a constraint in the way to digital finance. Sharon Buteau, Preethi Rao, Fabrizio Valenti (June, 2021), also stated the fact that there is a wide scope of financial digitalization in India as only up to 50% of Indian population out of those who are having mobile with internet connection are using digital finance. They have also depicted the fact that digital finance will be helpful to the middle and low-income level population in many ways like less transaction cost, less time consuming, more awareness about saving and investment, can be used in abnormal conditions and more safety. But there are some gaps that have been found in the study that are restricting digital finance in India, like it has been mentioned in the study that three vulnerable segments in India i.e., women, micro enterprises and domestic migrants. These vulnerable groups are having their own challenges with regards to technology, awareness and confidence level. This gap can be minimized by providing customized solutions. The researcher found that data privacy is also a measure bottleneck in the way to digital finance. Begum Mohammed Farzana (2019), has stated that the digital finance initiative will be helpful to create a connection among the different groups of the society and accelerate the financial inclusion with financial stability in India. But to achieve the mentioned objective there is an urgent need to spread education and awareness in remote areas and among the underprivileged population of India.

Rana, Luthra, Rao and H.R. (2020), have done research on challenges to digitalization of financial services in emerging economies. They have identified eighteen major challenges to digitalization of financial services with the help of interpretive structural modelling (ISM) and fuzzy MICMAC approach. There are two most important variables i.e., risk of using digital services and lack of trust is the main challenge in the way of digital finance followed by high cost and low return. Purva Khera,Stephanie Ng,Sumiko Ogawa,Ratna Sahay (2022), analyzed the development of digital financial inclusion in 52 emerging markets and developing economies. They found that to enhance financial inclusion, adoption of digitalization of financial services, policies with respect to digital finance are very much required. They have also suggested reducing the variation across the countries and regions for the enhancement of digital finance.

Raghavendra Nayak (2018), has pointed out that major issues in digitalization of financial services in Rural India in comparison to Urban India are due to the lack of literacy & financial literacy, lack of infrastructure facility (electricity, network issue & communication network), a smaller number of people using smartphones, lack of banking habit, resistance to new technology and cost. Due to all these challenges, the road to take the country from predominantly cash-based economy to predominantly digital economy will take time and needs to tackle many issues and challenges.

Growth of Digital Finance in India: Scope

The scope of digital finance is very wide and growing very rapidly as the Indian economy is a youth economy and today's youth is playing a significant role in making the Indian economy a digital finance economy. Vinay Kandpal and Rajat Mehrotra (2019), have depicted the fact in their paper that there is a wide scope of financial digitalization in India. They have mentioned in the paper that *"Fintech service firms are currently redefining the way companies and consumers conduct transactions on a daily basis. This is why global investments into Fintech ventures have been increasing at record speed – tripling to US$ 12.2 billion in 2014 from US$ 4.05 billion in 2013, and* *reaching US$ 19.1 billion in 2015. In India, the scale has been much smaller but at similar growth rates – investment in India's Fintech industry grew 282 percent between 2013 and 2014 and reached US$ 450 million in 2015."* According to their study, the Indian market has an untapped population (not using financial services) around 40% and 87% transactions are being made in cash. Small business is also lacking behind in using digital financial services, so there is a wide potential to enhance the digital finance and banking services in India. They have concluded their paper by stating the fact that to enhance the scope of digital finance in India financial institutions need to build customer's confidence and satisfaction with safety and privacy. Ravikumar T (2019), he has mentioned eight principles (GPFI, 2016) in his paper with which the scope of digital finance in India can be enhanced. Principles like to Promote a Digital Approach to Financial Inclusion, Balance innovation and risk to promote DFI, Provide legal and Regulatory framework for DFI, Expand DFS infrastructure ecosystem, Establish accountable digital financial practices to shield customers, Strengthen digital and financial literacy and awareness, Facilitate Customer Identification for Digital Financial Services and Track Digital Financial Inclusion Progress can be proved as a milestone in the way of digital finance.

Conclusion

Digitalization is the need of today's era. If we would like to survive in this rapidly changing world, we should adopt new technologies which are fast, efficient and sustainable. has launched Digital India flagship programme with a revelation to enable a digital strong economy in every aspect. Digitalization of finance, agriculture, manufacturing, railways, education, pharma services, aviation almost every segment of the society. During the year 2006 National e-Governance Plan (NeGP) was initiated in India to cover a large segment of services and work viz. financial services, agriculture, land records, health, education, passports, police, courts, municipalities, commercial taxes and treasuries etc. This initiative has got wings during the year 2014 when an emphasis has been given to financial inclusion. Just with a click on the mobile phone, laptop and computer without taking much time we can avail the financial services and complete the financial transactions. This chapter has focused on three important aspects of digital finance in India (i) Growth of Digital Finance in India, (ii) Challenges in Digital Finance in India and (iii) Scope of Growth of Digital Finance in India. This chapter depicted the fact that still there is a gap in the digital finance services and awareness about the same in India. This is the responsibility of academicians, researchers, policy makers, financial institutions and to eliminate this gap to make our country digital finance efficient. There is a need to empower underprivileged members of the society through literacy and financial literacy. Good infrastructure like internet connectivity and communication systems will be helpful to enhance digital finance in rural/remote areas.

References

Barik, R., & Sharma, P. (2019). Analyzing the progress and prospects of financial inclusion in India. Journal of Public Affairs, 19(4), e1948.

Begum, M. F. (2018). An Overview of Digital Financial Services in India: Concept, Initiatives and Advantages. Asian Journal of Management, 9(3), 1139-1144.

Buteau, S., Rao, P., & Valenti, F. (2021). Emerging insights from digital solutions in financial inclusion. CSI Transactions on ICT, 9(2), 105-114.

Derryl D'Silva, Zuzana Filková, Frank Packer and Siddharth Tiwari Monetary and Economic Department December 2019

Ghosh, C., & Hom Chaudhury, R. (2020). Determinants of digital finance in India. Innovation and Development, 1-20.

Kandpal, V., & Mehrotra, R. (2019). Financial inclusion: The role of Fintech and digital financial service

Khera, P., Ng, S., Ogawa, S., & Sahay, R. (2021). Measuring digital financial inclusion in emerging market and developing economies: a new index. Asian Economic Policy Review.

Nayak, R. (2018). A conceptual study on digitalization of banking-issues and challenges in rural India. International Journal of management, IT and Engineering, 8(6), 186-191.

Rana, N.P., Luthra, S. & Rao, H.R. (2020), "Key challenges to digital financial services in emerging economies: the Indian context", Information Technology & People, Vol. 33 No. 1, pp. 198-229. https://doi.org/10.1108/ITP-05-2018-0243

13

Digital Transformation and Digital Finance

Tinu Anand
Assistant Professor, Gitarattan International Business School

Dr. Savita
Assistant Professor, Gitarattan International Business School

Abstract

This article aims to present a comprehensive understanding of digitalization and how changes in technology affect every sector of the economy. Digital finance is the main topic of the current article. A wide range of cutting-edge financial goods, fresh ventures, financial software, and improved methods of customer relationship management and engagement are all part of digital finance and are offered by FinTech firms and forward-thinking financial service providers. With the advent of the fourth generation of telecommunications (4G) in 2016, the country's current trend of e-finance services has expanded at an alarming rate. It demonstrates the phenomenal expansion of the internet into every nook and cranny of Indian society, which has empowered both consumers and businesses by allowing them to cut transaction costs, process paperwork quickly online, and gain access to information right away. Online finance may significantly increase productivity for firms and cut expenses for internal operations including time and billing, labor management, and expense reporting. The Internet offers businesses the opportunity to engage in digital marketing and product promotion, giving them the ability to customize the online experience to meet the particular demands of each customer by utilizing AI (artificial Intelligence). The current study will concentrate on a few crucial aspects related to digital banking. In recent years, academics and scholars have realized the significance of digital technology and finance. Financial technology has various advantages. Banks benefit from digital finance, as do depositors from the efficiency and openness of the services they receive, and money market authorities also greatly benefit from it. Digital financial inclusion has the potential to enhance people's lives because it aims to touch base.

Keywords: *Digital Finance, FinTech, Digitalization, InsurTechs, WealthTechs*

Digitalization and Digital Finance

With the change in the world due to digitalization everywhere. Companies are forced to reinvest in their business model, simplify their current business operations and meet customer expectations thereby providing them more satisfaction.

Financial services provided via phones, computers, computers, and smart cards are termed digital finance. (Manyikaetal, 2016) [1] Gomber et. al .(2017),[2] define digital finance as an asset of a new range of financial products, innovative businesses, and software-based finance. The author also highlighted the new methods of consumer handling using technology-based relationship management systems-provided by FinTech organizations. As such, there is no fixed definition for digital finance but one can define it as a combination of products and services that are delivered by providers using resources like smartphones and the internet. There is no need to deal directly with the financial service provider for every transaction.

Digitalization can be defined as a process of integrating digital technologies into the working of business organizations for the ultimate purpose of improving processes.(Andersson et al., 2018; Gimpel et al., 2018)[3][4] is anDigital Technology is an integral part of any economy and society. (Bumann & Peter, 2019)[6] Stated that in wider terms Digital Transformation also includes additional terms such as digitization and digitalization. Digitalization is related to various fields, Digital Finance is one of the components.

According to the official website of the European Union, the term "digital finance" is used to describe how new technologies have affected the financial services sector. The old method of providing banking and financial services has been revolutionized by several new products, applications, processes, and business models. According to Capgemini, the major objectives of digital finance are to be smarter, easier, and safer.

Digital finance, according to "Gomber" et al. (2017),[7] includes a wide range of new financial products, businesses, software, and methods of client interaction and communication that are provided by FinTech firms and creative financial service providers.

Digital finance is a type of financial service offered through devices like smartphones, laptops, the internet, or cards that are connected to reputable digital payment systems. Digital finance has the potential to offer convenient, secure, and cost-effective banking services. With digital money, customers have more control over their finances, can make decisions quickly, and can send and receive payments.

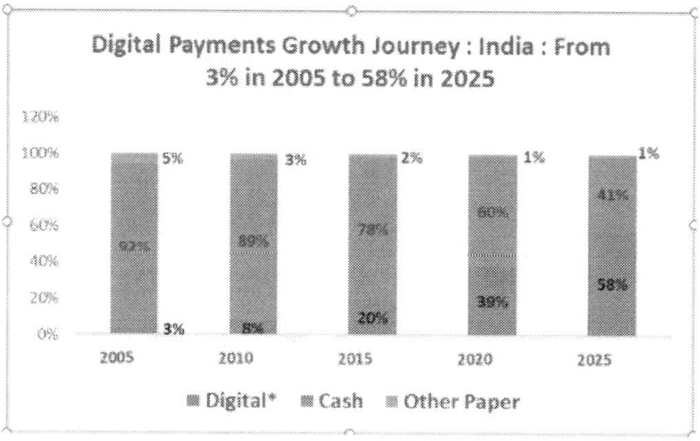

Source: #India #Digital Payments 2020 – Launching the First Adoption Index - Time is Now!

Figure 1

The growth of digital payments from 2005 to 2020 is depicted in figure 1 above, along with projections for future growth. Around the year 2000, India's digital payments adventure began, and it has since grown exponentially. By 2025, it is predicted that 58% of all cash and non-cash installments will be made using digital payment methods, which is a fantastic increase from the 3% in 2005.

Digital finance usage at a large scale offers a long list of benefits for service providers as well as for the customer. Digital finance may contribute to more and more financial inclusion in a nation and at the same time, it helps in the coverage area of financial services to the non-financial sector. The usage of mobile phones has increased dramatically, especially in developing nations, and using this demographic advantage, the reach of basic services can be extended to the individual level. (World Bank, 2014).[8]

Another reason to focus more on the usage of digital finance by developing countries is that it has the power to furnish inexpensive, appropriate safe banking service to lower sections of the society

(CGAP). Due to vast innovations in digital financial services at the global level, a major proportion of the total world's population can take advantage of developments in the area. At the same time, they can shift from cash-based transactions to technology (Internet & software) based transactions. (CGAP).

Various Benefits of Digital Finance, also encourage the respective bank clients to switch from offline banking to Online Banking Mode. Some of the encouraging facts in this regard are as the number of UPI transactions is increasing due to which many of the workforce and other people could be benefited from contactless payments during even crucial times. The heavy queue system can be avoided at Banks with the usage of Digital banking services.

Digital Financial Services

The group of financial services that can be accessed and provided through certain digital routes is referred to as Digital Financial Services (DFS). DFS are services that are offered and accessed by customers using their mobile devices, laptops, Point-of-Sale (POS), ATMs, etc. Here are a few examples of digital financial services:

1) **RTGS (Real Time Gross Settlement):** This financial service allows for the real-time and gross movement of money between banks.
2) **Retail Electronic Clearing Payment:** This includes NEFT, or National Electronic Fund Transfer, and IMPS, or the "instant payment system."
3) **Bank Cards:** In India, bank cards are the most popular form of electronic payment. Debit cards, credit cards, and prepaid cards are all included in this category.
4) **Unified Payment Interface (UPI):** Also referred to as UPI, this technology has become extremely widespread, particularly in urban areas. It integrates various bank accounts (from any partner banks) into a single mobile application, enabling a variety of banking functions, smooth fund routing, and merchant payments all under one roof. 224 banks are now operational on the UPI.
5) **Mobile wallets:** A mobile wallet, sometimes referred to as a digital wallet or e-wallet, is a kind of virtual wallet that can connect your debit card and/or credit card information on a mobile device to your mobile wallet software and allows you to send money online to the wallet.

6) **Point of Sale (POS) Terminals:** A point of sale is where transactions are completed. A POS could therefore be anywhere, such as a market, a city, or a shopping center. On a smaller scale, though, it is regarded as a place where a customer completes the transaction, similar to a checkout counter in a mall.

7) **Mobile Banking:** The majority of banks offer apps for smartphones and tablets that can be obtained from application stores like the Microsoft Store, Google Play Store, and App Store. With the use of a smartphone or tablet, it enables users to carry out various financial operations remotely.

Fintech

Financial technology means the incorporation of technology into the products and services delivery process by financial service providers for better and more effective services. In other words, one can say that it is an innovative technique for superior management of financial functioning, and processes by integrating the latest software technology and smartphones. It is convenient for all the parties i.e. business organizations and users. FinTech highlights the usage of a new business framework that offers more safety, adjustability, effectiveness, and scope as compared to traditional financial services. (Lee 2015a).[9]

The term FinTechsignifies the importance of innovators in the financial sector. These innovators may be in the form of new and upcoming organizations or there may be some already existing organizations that want to expand their market horizon. A few examples of Fintechs in the Indian structure are Paytm, Mobikwik, PolicyBazaar, etc. Lee (2015b)[10][11] provided two categories of innovators in FinTech. One is "Sustaining FinTech" and the second is "Disruptive FinTech".

Initially, the role of information technology is recognized in the finance sector only as an instrument for better performance. Nowadays, companies that are fully IT-supported are entering the financial sector. These companies are grabbing the customers of traditional financial services providers. The main reason for the inclination of even loyal customers towards Fin techs is that they are providing better and innovative solutions to customers' requirements.

The main model of FinTech companies focuses on inexpensive and internet-based working to grab the market share of popular service providers. As per the opinion of experts, in the future banks will be doing the job of collecting money from the customers, rest of the services will be totally under the control of FinTech companies.[12]

As per CFA Institute "Fintech, or financial technology, refers to the technological innovation in the design and delivery of financial services and products. Technology in finance continues to evolve; advancements include the use of Big Data, artificial intelligence (AI), and machine learning to evaluate investment opportunities, optimize portfolios, and mitigate risks."[13]

Fintech is not a new industry; it is only one that has developed quickly. The financial world has, in some ways, always relied on innovation, whether it was the introduction of Visa cards in the 1950s or the emergence of ATMs, electronic trading floors, personal accounting software, and high-frequency trading in the years that followed.

As per Forbes Advisor, "Fintech refers to software, algorithms, and applications for both computers- and mobile-based tools. In some cases, it includes hardware, too—like smart, connected piggy banks or virtual reality (VR) trading platforms. Fintech platforms enable run-of-the-mill tasks like depositing checks, moving money among accounts, paying bills, or applying for financial aid. They also encompass technically intricate concepts like peer-to-peer lending or crypto exchanges."[14]

The driving forces behind financial innovation vary from project to project and application to application. However, the most recent developments use information science, blockchain, and AI computations to handle everything from managing mutual funds to interacting with credit risk. The field of administrative innovation known as "regtech" is currently being studied to better understand the complex world of consistency and administrative challenges faced by businesses such as, you guessed it, fintech.

Concerns about online security in the fintech industry have grown along with the technology. The massive growth of fintech businesses and commercial hubs on a global scale has led to an increased awareness of the foundation's flaws and made it a target for cyberattacks. Fortunately, innovation continues to advance to reduce risks that are constantly emerging and limit current extortion bets.

Transformation of Banking and Finance into Digital Form

Banks and other financial institutions are stepping up their efforts to digitally transform, investing more money and resources in projects like enhanced mobile offerings, data analytics, and artificial intelligence (AI) to not only hyper-personalize consumer banking and wealth management experiences but also to gain an advantage in a market that is becoming more and more attractive to Big Tech competitors.

To provide customers with more seamless experiences and take market share away from well-established financial institutions, big digital corporations are looking to join a variety of fintech, insurance, wealth tech, and other financial services start-ups (FIs). Fintech startups and conventional FIs are both looking to innovate in all facets of the digital banking space since the epidemic drove many consumers toward digital experiences, such as digital banking and contactless payments. Existing banks must either collaborate with their new, tech-savvy rivals or update their antiquated systems if they want to stand a chance of competing in this new environment.

Fintech Revolution

In recent years, India's Fintech sector has experienced a remarkable rise in funding; predictions for 2021 that are expected to be worth more than $8 billion have been made.

While Alternative Finance and payments fragments comprised more than 90% of the area's speculation streams in 2015, there has been a significant shift towards a more even handed dispersion of venture across areas including InsurTechs, WealthTechs, and so forth.

There are around 23 Fintechs in India that gained "Unicorn Status" out of which one-fifth of Unicorn startups are from Fintech.[15]

The no. of real-time transactions in India exceeded 48 billion in 2021, surpassing the combined volume of the world's top five economies (the United States, Canada, the United Kingdom, France, and Germany) by 6.5 times. This led to cost savings of USD 12.6 billion for Indian businesses and consumers in 2021.

By 2025, the digital investment market is expected to reach a value of $14.3 billion, up from $6.4 billion in 2021, with a 5-year CAGR of 22.4%.

The market for digital payments in India is predicted to more than treble from $3 trillion now to $10 trillion by 2026. Digital payments (non-cash) will account for over 65% of all payments by 2026 as a result of this exceptional rise, or 2 out of 3 transactions (by value).

The culmination of years of work establishing the framework for creating key enablers through significant projects is India's Fintech revolution:

- **Jan DhanYojana:** Jan DhanYojana: The world's biggest monetary consideration drive, "Jan DhanYojna", has assisted in new keep money with bookkeeping enlistment of more than 450 Mn recipients for direct advantages move and openness to a large group of monetary administrations applications like settlements, credit, protection, and benefits empowering FinTech players to fabricate innovation items to enter the enormous buyer base in India.
- **Financial Literacy:** The establishment of the National Center for Finance Education and the execution of the Center for Financial Literacy project by the Reserve Bank of India are two recent initiatives in India aimed at promoting financial literacy. With the help of these methods, financial education is anticipated to improve for all demographic groups in India.
- **E-RUPI:** e-RUPI is an individual and reason explicit computerized installments instrument to consider contactless and credit-only installment arrangements and will play a vital role in strengthening and ensuring the consistency of the Direct Benefits Transfer. For the Covid-19 vaccination, the arrangement is being made for credit-only installments.
- **"India Stack":** India Stack is a bunch of APIs that permits states, organizations, new businesses, and interesting computerized and interesting computerized Infrastructure to tackle India's difficult issues towards presence-less, paperless, and credit-only help conveyance. "The India Stack has been the main thrust behind the sped-up advancement of Fintechs". It is one of the main computerized drives embraced universally, pointed toward setting up a public advanced framework given open APIs to advance public and confidential advanced drives and plays that had a reactant impact in India's computerized establishment and development.

Top Fintech Unicorn Companies in India

- Paytm
- Rozarpay
- Pine labs
- Groww
- Bharat Pe
- Policybazar
- CRED
- Zeta

- CoinSwitch
- ChargeBee Technologies
- Vedantu And Many More

The most unicorn-status startups have been found in the financial and insurance sectors. Payments, loans, wealth technology, insurance technology, and regulation technology are some of the sub-segments that make up the Indian fintech ecosystem. 70% of Fintech Unicorns were involved in payments and digital lending, according to a survey by Credit Suisse on start-ups and FinTechs in India. According to Razorpay, over the previous 12 months, digital payment transactions have increased by up to 76% with several new customers. India has become a refuge for digital payments thanks to its friendly governmental environment, legislation, and UPI.

INDIA – A FinTech Superpower Globally

The world's greatest rate of FinTech adoption is in India. There are around 6,636 FinTech startups in India and it has one of the fastest-growing Fintech sectors in the world.

Market size for the Indian "FinTech" sector was likely to reach $50 billion in 2021 and $150 billion by 2025. In terms of digital payments by volume is a CAGR of 50% and a value of CAGR of 6%, India's payment landscape has grown over the past ten years to become the most sophisticated payment system. The value of Fintech transactions is anticipated to increase from US$ 66 billion in 2019 to US$ 138 billion in 2023, at a CAGR of 20 percent.

"Payments, Lending, Wealth Technology (WealthTech), Personal Finance Management, Insurance Technology (InsurTech), Regulation Technology (RegTech), and many other subsegments make up the Indian Fintech business ecosystem".

In the financial year 2022, India's fintech industry received $8.53 billion in financing (in 278 deals). 323 banks had signed up to use the UPI i.e. Unified payment interface of India as of May 2022, and the system has processed 5.9 billion monthly transactions totaling more than $130 billion. India has 23 fintech firms that, as of June 2022, have achieved "Unicorn Status" with a 1 Billion dollar valuation.

Future Scope of Digitalisation

Even though there is a danger associated with e-banking and e-finance services, e-Finance is growing in India at a significant rate compared to e-commerce. ICT developments like artificial intelligence and machine learning have made it possible to prevent stable economic growth through financial inclusion throughout the nation. Financial institutions are improving their procedures as a result. Procedures and frameworks for managing fraud risk to decrease the likelihood of fraud and shorten the time required for detection. However, funding for attempts to combat fraud continues to compete with other corporate activities and is most challenged in terms of cost and benefit. Agencies for regulation and investigation are working to adapt to the new environment.

Conclusion

Although the financial sector has historically been an early adopter and heavy user of breakthroughs in ICT, the introduction of creative business models and the entrance of new competitors have had a significant impact on the dynamics of the industry today. The incumbent suppliers of banking and financial services are undergoing a significant transformation as a result of the quickly shifting economic climate as well as onerous and expensive regulatory requirements: digital finance.

Traditional service providers' business models are under threat from new technologies, more tech-savvy customers across generations, and the digitalization of the sector. In all pertinent business operations in the financial services industry, one can notice both novel forms of disintermediation and new rivals. In light of this, the majority of market participants work to develop specialized, intelligent, adaptable, yet cost-effective financial goods and services and aim to reach new heights of client centricity.

The current essay advances the new FinTech revolution by giving an outline of digitalization and digital finance. Particularly after the launch of Digital India, digital finance has taken on increased significance in the Indian financial sector. The Indian financial sector's many industries are all-embracing fintech. The adoption rate of fintech worldwide is highest in India. In India there are around 6,636 startups, which has one of the fastest-growing Fintech sectors in the world.

In India, there are about 23 fintech businesses that have achieved "unicorn Status." There are various drawbacks to digital finance, including its cost, security, adaptability, etc. Every human being wants to use digital finance in their daily lives.

References

Hussain, H. B., Darzi, M. A., Bhat, S. A., & Bhat, A. R. (2019). Financial Inclusion and Digital Finance. Sankalpa, 9(2), 6-11.

Gomber, P., Koch, J. A., & Siering, M. (2017). Digital Finance and FinTech: current research and future research directions. Journal of Business Economics, 87(5), 537-580.

Andersson, P., Mattsson, L. G., Movin, S., Mähring, M., Teigland, R., & Wennberg, K. (2018). Digital transformation supporting public service innovation: Business model challenges and sustainable development opportunities. Managing digital transformation, 217-243.

Gimpel, H., Hosseini, S., Huber, R., Probst, L., Röglinger, M., & Faisst, U. (2018). Structuring digital transformation: a framework of action fields and its application at ZEISS. *Journal of Information Technology Theory and Application (JITTA)*, *19*(1), 3.

Gimpel, H., Rau, D., &Röglinger, M. (2018). Understanding FinTech start-ups–a taxonomy of consumer-oriented service offerings. *Electronic Markets*, *28*(3), 245-264.

Bumann, J., & Peter, M. K. (2019). Action fields of digital transformation–a review and comparative analysis of digital transformation maturity models and frameworks. Digitalisierung und andere Innovationsformen im Management, 2, 13-40.

Gomber, P., Koch, J. A., &Siering, M. (2017). Digital Finance and FinTech: current research and future research directions. *Journal of Business Economics*, *87*(5), 537-580.

World Bank (2014) Digital Finance: Empowering the Poor via New Technologies, April 10. Available at: http://www.worldbank.org/en/news/feature/2014/04/10/digital-finance-empowering-poor- new-technologies Accessed on 10 November 2017.

Lee P (2015a) The fintech entrepreneurs aiming to reinvent finance. Euromoney (UK) 46(552):42–48.

Lee, D. K. C. (2017). Fintech tsunami: Blockchain as the driver of the fourth industrial revolution. Available at SSRN 2998093.

Lee, D. K. C., & Low, L. (2018). *Inclusive fintech: blockchain, cryptocurrency, and ICO*. World Scientific.

Abbasov, A. M., Mamedov, Z. F., &Aliev, S. A. (2019). Digitalization of the banking sector: new challenges and prospects. *Economics and Management*, (6), 81-89.

https://www.cfainstitute.org/en/research/fintech.

https://www.forbes.com/sites/forbesbusinesscouncil/2020/10/15/digital-transformation-and-its-impact-on-finance-and-the-banking-system/

https://blog.cfte.education/fintech-unicorns-in-india-full-list/

14

Effect of Digitisation on Indian Financial Markets

Pranav Sharma
Research Scholar, School of Management & Commerce, Manav Rachna University

Dr. Shweta Goel
Associate Professor, School of Management & Commerce, Manav Rachna University

Abstract

The technology emergence has had a rapid and irreversible impact on the world and the Indian securities market has also witnessed these changes. In the last twenty years, many individuals have been connected to the internet.. From the nineteenth century, the capital market of India made a comeback. By Eliminating additional customer and geographic barriers, the Internet has created achievable monetary products and services. The first investors were single killed by their brokers but today they are participating more in buying and Commercialization of shares with the help of the internet. IT has saved time, energy and Cash by providing access to the market from anywhere at any time. Here in this paper we will discuss the commercialization of the internet in India and the scope of online commercialism in India.

Keyword: *Digitization , Internet, Financial Markets JEL Classification; E44, L86*

Introduction

The world is going computerized, and the pace of change is rising. India has approx 300 millions internet clients. Inside this, Google gauges it took 20 years, a long time for India to score its first 100 mn clients, whereas the following 100 mn took 2 years a long time and 1.3 a long time individually. This is expected to reach 600 mn by 2020, with clients over sex and age-groups. Comscore's Jan 2022 figures may be more pertinent for money-related administrations. It gauges approx 200 mn Males of age 25+ a long time gone by monetary websites. Whereas money-related administrations are not a Male-domain in any sense, the supreme numbers are eye-catching by themselves.

Going digital really mean business?

Advanced strategies incorporate company destinations, outside entries, social-media, smartphone applications, web-chat, IVR phone office, nearness on web-aggregators, SMS, email, etc. These can be Own-Media, Paid-Media or Earned-Media. Earned media implies surveys, notices, posts and offers which clients make voluntarily on social-media resources. Paid media implies paid advertisements/ promotions. In terms of gadgets, smartphones are the biggest and speediest developing medium. Tablets are also developing. Computers are lessening in relevance. The reason for computerized strategies is to inform, lock in and execute clients/prospects with their commerce. It began as an 'addon', but has moved to a 'must-have'. In budgetary administrations, it has become an important distribution channel. The objective for money-related item makers is to diminish the costs as well as increment incomes. For money related item operators, it is truly to extend the revenues.

Digitization should be used as a backbone or for support?

The monetary administrations commerce is customer-facing, competitive, dissemination- sensitive and turnaround-time touchy. In case the company's target universe is progressively going digital, it makes sense to embrace advances such as the Spine, instead of a Back. This may give it a 'first-mover advantage', which can have bearing on its victory in an advertisement like India, which is still advancing in terms of modern monetary items. It moreover depends on the industry it is in. For case: it can be a back in insurance but it may be a must in broking.

But first-mover advantage requires not continuously interpreted into brand stickiness. For that, digital strategies have to give incredible involvement to clients over ease of use, substance and access. Nowadays, clients are not so focused with brands, as much as with convenience and esteem. Firms too need to go to the next-level by making highlights that compel repeat-visits and client stickiness.

Areas Where Digitalization is Making Impact in the Value Chain

From Customer Engagement to Customer Fulfillment

- **Buying Decision and completion :** Computerized strategies have been to a great extent utilized for information-access, item mindfulness and brand nearness. In comparison, actual buying action was moderate. Reasons to incline toward offline buying can be many – a chance to have a conversation with the sales representative some time recently, to test the product physically, security concerns in almost all online installment doors, etc. However, with security coding getting to be exacting, clients are creating the comfort to purchase and total the exchange online. For illustration: Kotak's Banking and ICICI's Stash are cases of utilizing social-media/ digital platforms to totally manage an account exchange. Another challenge is to impact the want to purchase. This incorporates devices that help clients discover more profound bits of knowledge from the computerized sources which they might have otherwise got offline. For case: E-commerce firms put pictures of models wearing the articles of clothing, rather than putting the garment's picture. This helps clients visualize how they would see wearing it, and consequently hurries the decision to purchase. In monetary administrations, this incorporates expository devices to survey the product under scenarios/variables, which would have something else inquired the advisor personally.
- **Engaging with Existing Clients:** advanced strategies allow clients to purchase the product. Nowadays, item data is accessible on different stages. Hence, product data may not continuously be the most prerequisite of clients when they visit the company's computerized stages, but the get to the item for buying would be. For illustration: clients may compare data on shared funds on third-party web-aggregator stages instead of on the fund's stage, but may visit the fund's stage when they need to purchase specifically.
- **Lead Generation**: Another high-impact zone, advanced strategies offer assistance to generate audience interest through online challenges, publicizing or online apparatuses. In social-media, these are demonstrated strategies that inspire higher levels of audience engagement. This group of onlookers' interest makes a difference in coordinate contact with leads,

with whom the company can follow-up with a more focused on pitch. The conversion of group of onlookers support into lead era could be a key metric that helps companies evaluate if their computerized showcasing endeavors are moving within the right direction or not

- **Prospect Engagement:** This can be a high-impact zone where computerized strategies offer assistance in brand-building, information-sharing and develop the mindfulness of the products, with the point of changing over them to purchase. In money related administrations where the current penetration levels in India are still low due to need of mindfulness, this frame of engagement too makes a difference to inform the broader population why they got to save and contribute within the to begin with. Thus, it makes a difference to educate approximately the concept itself, even before educating almost the company's particular items. Given the tall level of competition within the budgetary administrations space, usually too a basic movement that creates top-of-the-mind brand recall once the prospect truly needs to convert. Ideally, he will reach out to either of the best 3-4 brands that come up immediately in his intellect. In conclusion, the core is additionally to supply stages which are tall on usability, content and comfort, since that can too affect the extreme change.
- **New Client Acquisition:** A branch of the prior "Lead Generation" point, while computerized media is getting to be significant as a channel of unused client procurement, it has not taken over the conventional media. By the way, it'll become more relevant as advanced appropriation develops. One range where advanced can play a major role in client securing is in recognizing target geographies where prospects reside. Since IP Address following is conceivable in advanced media, the interest from the audience can be traced to their area. In this way, companies can recognize specific geographies where they have to be aggressive in client procurement endeavors, instead of investing its rare assets over the complete nation.
- **Product Promotion & Comparison:** A high-impact region wherein company-owned and external-owned stages grant access to item data. This includes product-comparisons, where clients can straightforwardly compare product features. The product-comparison device may be a strategy to impact clients towards the decision to purchase. Policybazaar.

com is a case where one can compare protection policies. Moneycontrol and Esteem Investigate are stages which permit comparison of mutual reserves. Company-owned social-media is a fast-growing medium to promote mindfulness of modern items, complementing conventional media like TV. Social-media is presently getting to be a space where individuals review brands/companies some time recently. The reactions to offer assistance companies gauge the reaction for new/ existing items, based on the sort of audits it received, volume of interaction it created or reactions to any online challenge it conducted.

- **Preference-storage:** This implies getting to the platform over gadgets so that engagement isn't limited by users' location. For illustration: broking exchanging stages are presently open over all gadgets, be it computers, smartphones and tablets, so that clients can get financial information from any put and put exchange in case they need to through their device. This too incorporates putting away the client's preferences/favorites over gadgets to make future purchases speedier. Google browser permits clients to exchange their browser related preferences/favorites over gadgets, to help comfort. Stash, a smartphone app permits clients to store web pages, counting the substance, so that users can peruse it when they have to save time (whereas commuting, etc).
- **Data Collection for Managements' decision-making:** Trade Insights is all set to be a high-impact region where advanced strategies can revolutionize the way company administration utilizes information to require more choices that are educated. Huge volumes of significant client information can be captured and prepared, which suggests more relevant decision-making since more information is accessible to legitimize the inferences. This envelopes both online and offline engagement data, hence a larger universe of addressable clients. Enormous Information could be a center of commerce intelligence. It is still not being utilized by companies to its fullest degree, but is set to end up a key zone of computerized administration in a long time to come.
- **Reporting and Notifications:** This is often a must-have for money related firms to ensure ease-of-access and quick reaction. Occasional portfolio reports, account statements, transaction affirmations, installment alarms, installment affirmations, courier alerts, etc are all passed on to clients

immediately utilizing advanced mediums like emails, SMS and online portfolio apparatuses. In numerous cases, the clients themselves have access to the stage for getting to this information, either online portfolio stages or by sending an inquiry SMS to a predefined number. In this way, the possession actually transfers to the client which includes comfort and speed. For illustration: Motilal Oswal's Industry-first Online Account Opening was an office to form unused client accounts where the company was not physically displayed. The online platform educated clients on the status of the application until its completion. The challenge is that the client ought to have the comfort-level to use those computerized strategies.

- **Client Servicing and Client Feedback:** This is often another must-have for all businesses, not fair budgetary administrations. Computerized strategies of client adjusting help track the status of client inquiries and diminish turnaround-times and fetch of servicing. Since it is simple to determine the adequacy, this plays a key part in assessing client fulfillment and the requirement for any advancements. For illustration: many websites consolidate web-chat administrations where client inquiries are addressed immediately. Numerous social-media destinations permit enlisting client complaints which are taken up quickly. IVR phone offices permit coordinated exchange to the relevant division which the client's inquiry relates to.

Computerized stages make it straightforward to assemble criticism, which makes a difference to check client satisfaction. But online input shapes are kept as a choice in most cases, and clients tend to fill these as it were in the event that they have a negative involvement. Product-reviews section in destinations is valuable to type in criticism. Future clients can examine audits when exploring, and companies can make advancements based on it. Computerized feedback via social- media can moreover be a 'double-edged' sword on the off chance that the validity of the audience is moo, since their reactions will lead to erroneous inductions.

Literature Review

Brad M. Barber and Adaire Morse (2022) analyzed maximum financial specialists who switched from phone-based to online exchanging Those who exchanged to online trading perform well earlier to going online, beating the advertise by more than 2% yearly. After going online, they exchange more effectively, more hypothetically, and less productively than before—lagging the showcase by more than 3% yearly.

Decreases in advertised grindings (lower exchanging costs, progressed execution speed, and greater ease of getting to) don't clarify these discoveries. Overconfidence—augmented by self attribution predisposition and the dreams of information and control—can clarify the increase in exchanging and diminishment in execution of online financial specialists. They found that investors who select to create investments online are superior entertainers than those who don't go online some time after the switch but more regrettable entertainers after the move. The thought is that overconfidence actuates them to switch but at that point intemperate exchanging after the switch dissipates their benefits.

Haroun Alryalat, Yogesh Kumar Dwivedi, Jasna Kuljis, and Beam J. Paul (2006) analyzed the impact of online and conventional exchanging on viable advertisement execution on the NASDAQ. The reason for this paper was to display a basic examination on the competition between online (ECN) and conventional (Advertise Creator (MM)) exchanging on the NASDAQ stock trade. Online stock exchanging components at the trades are regularly a cross breed of dealers and sell-off markets. Distinctive viewpoints of exchanging execution, which is the most commonly utilized advertisement in the middle of display, were analyzed. This leads to a talk on: (1) the way that executes arrange is organized and (2) its effect on the viable market performance, exchanging fetched and financial specialist behavior.

Within the Indian setting, online exchanging can be properly called as a later marvel, and even today online exchanging isn't much well known among speculators for which a list of factors can be faulted. This reality is clearer from the data available where numbers of stock trades in India have developed from 7 trades in 1946 to add up individuals and others to 2.014 million. Over the past two long times, the esteem of all exchanges executed through the web on NSE has grown at a very high pace.

Analytical Discussion

Critical aspects to consider when an organization embarks on its Digital journey... Marketing the platforms across channels:

- Indeed on the off chance that one includes a extraordinary item, it may fall flat due to need of marketing
- Showcasing makes a difference make client engagement and cooperation, a key source to generate leads and change over clients to buy
- To advertise the platforms/features over gadget types

Investing into Talent:

- IT group incl. designers, coders, support and designers
- Look motor specialists (SEM/SEO) to track and monitor
- Marketers who exhort on what to say, where to say, when to say
- Specialists who bridge the crevice between space & specialized teams
- Security experts to guarantee information and user-info is secure

Investing into Content:

- Substance journalists to write/edit pertinent substance sections
- It has got to be made into Pieces, to form it helpful to consume
- Should be visual and intuitively pieces, because it upgrades engagement
- It should be upgraded and churned frequently

Buying vs Making decisions for Infrastructure:

- Making' makes sense on the off chance that computerized is to end up the backbone
- Indeed in case of merchants, Exclusive-Agreements can be entered into
- Can purchase source-code of stages to hold control eventually
- Upgradation is speedier in case done in-house instead of through vendor
- With cloud-computing, one can buy server space, rather than buying

Global Context

Monetary markets and administration committees are talking about how digitisation is transforming our society and how companies have to adapt. After a major effect, regularly broadly secured within the media, in particular segments such as the hotel industry affected by the entry of Airbnb, and taxis by that of Uber, digitisation is now moving toward a more crucial change of our economy.

Conventional trade is beneath weight taking after the gigantic development of e-commerce:

- Nourishment retailers are profoundly concerned around the entry of Amazon into their market (further to its procurement of Entirety Foods);
- Banks are stressed that their advertising share will be eaten up by the Fintechs;
- The media must adjust the administrations they offer;
- Car producers are seeing their industry created in reaction to innovations made by innovation mammoths and the coming changes in client behavior.

In short, many industry players are concerned by this digital disruption, a lot more it seems than just hotel owners and taxi drivers. These technological innovations are known as "disruptive". They imply a dislocation of the existing model and therefore disturb the markets of traditional businesses. Joseph Schumpeter already postulated this concept in the last century when he spoke about creative destruction. At the current stage, how has the stock market integrated the impacts of these disruptive technologies?

Conclusion

The Winners

Because it shows up very self-evident nowadays, the enormous champions are the tech stocks, in specific, the famous "FAANG" (Facebook, Apple, Amazon, Netflix, and Google). These five companies by themselves speak to over 10% of the S&P500's advertise capitalisation and nearly 25% of its execution since the beginning of the year. The enthusiasm is effective, as often as possible defended by ever-increasing profit, but today's sector valuations can be of concern as they appear moderately exaggerated (on this point, see: Are US Values Truly Exaggerated?). The development potential of tech stocks appears more limited nowadays.

The Losers

It is less talked about, but digitisation has as of now made a number of casualties for which the technological transformation is bringing enduring weight to bear on their future profitability.

Within the S&P 500, two categories of companies stand out by their miserable performance. Firstly, certain exercises connected to the conclusion that customers (shops, department stores, media, and nourishment retail) have been rebuffed, accentuating the need for investor confidence in their commerce. Whereas most businesses are right now at the beat of their earnings cycle and are exchanging at their highs, the mark-down connected to these activities suggests that financial specialists have as of now coordinated the negative impacts of digitisation.

Furthermore, the markets moreover deserted oil and gas stocks. Here as well, the development of a new innovation - shale gas - clarifies the troubles being confronted by the division, however these are completely disconnected to the digitisation of the economy.

For other segments, speculators at this organization don't show up to be really concerned about the disturbance being caused by the economy's digitisation. The computerized wave will not necessarily be a tidal wave for all businesses. A few are able to adjust to the new environment and hence are able to stand up to it. They can alter their working strategies and refocus their procedure. Specifically, they can develop their e-commerce commerce and integrate unused innovation devices. Besides, certain trade divisions are secured by financial or administrative obstructions to passage that are regularly troublesome to overcome for new entrants. For illustration, mechanical exercises involve tall settled costs. Healthcare and the financial segment are entirely controlled . The digitization of the economy will clearly not have such destructive impacts for everyone, but maybe it is just a matter of time.

References

Sameer Kamath "Impact of digitization on financial markets" MOFSL Blog , 20-Jan-2017

Sameer Kochhar, R Chandrashekhar, K C Chakrabarty, Deepak B Phatak, eds., "Financial Inclusion", Academic Foundation, New Delhi, 2009

Barber, B. And Odean (2002) T., "Online Investors: Do the Slow Die First?" Review of Financial Studies, Vol. 15, 455–88.

http://www.mckinsey.com/mgi/publications/internet_matters/index.asp accessed on10/08/2011

Barber, B. And Odean (2002) T., "Online Investors: Do the Slow Die First?" Review of Financial Studies, Vol. 15, 455–88.

https://www.nseindia.com/

http://www.livemint.com/2011/07/03233438/Let-Internet-have-a-greaterro.html?h=B accessed on 07/07/2011

Desai Vasant (2010), "Financial Markets and Financial Services", 212-228

www.strategyand.pwc.com/media/file/Strategyand _Maximisi ng-the-Impact13. www.thediplomat.com/2016/12/indias demonetization-time-for-digital Development (IJTSRD) ISSN: 2456-6470 | IF: 4.101 | Oct 2018 Page: 28 www.strategyand.pwc.com/media/file/Strategyand -of-Digitization.pdf

❑❑❑

SECTION - 5
Higher Education Sector

15

Adoption of E-Learning Resources in Higher Education Institutions: A Systematic Literature Review

Mona Sharma
Manav Rachna University, Faridabad

Abstract

E-learning is increasingly being used and adopted in the higher education institutes for enhancing professional growth and development. In education, the use of e-learning is mainly used as a resource that supports teaching. For this new advancement in universities, this paper explores the factors which are behind the e-learning adoption. This study reviews literature and provides a scholarly background to the study by including some contributions given by various researchers as well as institutions on e-learning acceptance in higher education. 25 articles from 2000 to 2022, were extracted through systematic procedure of search, and exclusion and inclusion criteria were adopted for the study. The paper also identifies the factors of willingness in students and faculties in adopting e-learning technology and resources, challenges to adopt e- learning in universities, models that have been used to support the study, findings in the studies to support the research. Results from a systematic review of 25 studies, revealed key determinants of e-learning adoption. This paper also provides different suggestions for effective implementation of e-learning in higher education institutions.

Keywords: *E-learning, Adoption, Higher Education, Online Learning, Universities.*

Introduction: Concept of E-Learning

With the continuous advancement of information and communication technology resources and tools, ICT has transformed all the aspects of teaching and learning. There is a drastic change in the teaching learning system in higher education after COVID pandemic. There is a drastic change in how people communicate with each other either at personal level or inter-organizational communication, wherein there are no regional boundaries (Memon & Meyer, 2017). ICT has resulted in significant change in teaching-learning environment which have a positive impact on students' and faculties' performance at personal as well as organizational level. The frequent use of the Internet and the mass distribution access to the advanced technologies were the main factors for the evolution of remote teaching services. Research identifies that the e-learning resources adoption , enhances the students' pace of learning, helps the students with advanced techniques, helps the faculties in their training and development, makes the course plan more interesting, and make the students prepared for the global competitive education environment (Sayaf et al., 2021).

The Internet is one of the best means to give and obtain information regarding anything. Students have grown up using the technologies since childhood like video games, cell phones, computers, that is why they are also called "digital natives" as mentioned by (Prensky 2001). E-learning is integrated learning assisted by different applications, processes, methods of teaching, learning platforms like Youtube, zoom etc.

Meaning of E-Learning

E-learning can be defined as how knowledge can be acquired through computers. In the last ten years, e- learning has become a requirement in educational institutions because of the benefits that it offers to the students and teachers. Students and users make use of network technologies in sitting in different locations to interact with other students. The E-learning method is very flexible for the users, learners and instructors at various timings.

Students can study anytime anywhere as they want. This paper tries to identify how the students who are studying in a higher education institution and faculties who are working there, react while they adopt the e-learning resources. Also, there are certain factors which are behind the acceptance of e-learning by the users. The factors that will be the reason behind the satisfaction among the students and faculties for acceptance of e-learning, may help the administrators and policy makers in understanding their behaviour so that they can implement the same in the organisation (Grandon, Allshare & Kwan, 2005)

Abbad et al (2009) defined e-learning as any learning which is done electronically. In other words, e-learning is basically defined as digital learning which is assisted with technology-facilitated digital devices such as cameras, online software, powerpoint presentations, projectors, videos, tablets, OHP, various operating systems, personal computers which help in the communication between the students and faculties (Markus and Robey, 1998). E-learning also makes the teaching learning process very easy to implement due to its usage and convenience. It also includes various learning applications which allow the participants to teach and learn from distance and also allow them to interact face to face in an online mode (Markus and Robey, 1998). The e-learning adoption is believed to bring various advantages to both students as well as institutions. From institutions' point of view, e-learning improves observing students' ability, makes easy transfer of learning content and also reduce costs e.g. travelling cost, building cost etc.

The study's research aims are as follows:

- What are the factors behind the adoption of e-learning by students and faculties in higher education institutions?
- What are the suggestions for enhancing the e-learning adoption in higher education for effective teaching and learning?

Methodology

This study is based on literature review. We have studied articles from some reputable database journals such as: IEEE explore, EBSCO, ProQuest and Google Scholar. We applied "E-learning Adoption in higher education" as a keyword. We found 60 articles, after reviewing the article's abstracts we have chosen 25 relevant papers. We have done the analysis and the results and findings will be discussed in the discussion part below. Research papers published in high profile journals, review articles and conference proceedings whose citation indexes were highly evaluated.

Master thesis were not included in this study and opinion research papers were also excluded. Individual analysis was also done to evaluate articles that did not suit the methodology of this study. To strengthen the review's validity and reliability only peer reviewed articles and conference papers were taken into consideration and dissertations, books , volumes were ignored.

To abstract papers from databases terms like E-learning adoption, online learning, M-learning were used.

(Tableb1) displays 25 papers that were selected for the study. The first paper which was published in 2022 was related to the different advantages that e-learning offers and also the challenges faced by users in higher education institutions . The second paper was into the domain of constraints and factors that are responsible for e-learning adoption during COVID-19 pandemic. Other papers were also related to the models that have been used to evaluate the factors of adoption of e-learning in higher education institutions .

Table 1: Review of papers selected

TITLE OF THE PAPER	AUTHORS	PUBLICATION YEAR	SOURCE TITLE
Electronic Learning Benefits And Challenges In Malawi's Higher Education: A Literature Review	Limbani Chrispin Gama, George Theodore Chipeta, Winner Dominic Chawinga	2022	Education And Information Technologies
Constraints Of Transition To Online Distance Learning In Higher Education Institutions During COVID-19 In Developing Countries: A Systematic Review	Placidius Ndibalema	2022	E-Learning And Digital Media
Teachers' Digital Competencies In Higher Education: A Systematic Literature Review	Veronica Basilotta-Gomez-Pablos1 , Maria Matarranz , Luis-Alberto Casado-Aranda And Ana Otto	2022	International Journal Of Educational Technology In Higher Education
Blended Learning Adoption And Implementation In Higher Education: A Theoretical And Systematic Review	Bokolo Anthony Jr,Adzhar Kamaludin, Awanis Romli,Anis Farihan Mat Rafei, Danakorn Nincarean A. L. Eh Phon, Aziman Abdullah,-Gan Leong Ming	2022	Technology, Knowledge And Learning

TITLE OF THE PAPER	AUTHORS	PUBLICATION YEAR	SOURCE TITLE
Artificial Intelligence And Big Data: The Advent Of New Pedagogy In The Adaptive E-Learning System In The Higher Educational Institutions Of Saudi Arabia	Mohammed Arshad Khan , Maysoon Khojah, And Vivek	2022	Hindawi Education Research International
A Literature Review: Efficacy Of Online Learning Courses For Higher Education Institution Using Meta-Analysis	Mayleen Dorcas ,B. Castro And Gilbert M. Tumibay	2021	Education And Information Technologies
Exploring Factors Influencing Mobile Learning In Higher Education – A Systematic Review	Yuvika Gupta, Farheen Mujeeb Khan, Sonal Agarwal	2021	International Journal Of Interactive Mobile Technologies
Factors Influencing E-Learning Adoption In India: Learners' Perspective	P S Vanitha,Sreejith Alathur	2021	Education And Information Technologies
Adoption Of E-Learning At Higher Education Institutions: A Systematic Literature Review	Rushna Khalil Awan, Gul Afshan , Atia Bano Memon	2021	Multidisciplinary Journal For Education ,Social And Technological Sciences
Identifying Key Factors Affecting College Students' Adoption Of The E-Learning System In Mandatory Blended Learning Environments	Zhaoli Zhang, Taihe Cao, Jiangbo Shu & Hai Liu	2020	Interactive Learning Environments
Factors Affecting Student's Perceived Readiness On Abrupt Distance Learning Adoption: Indonesian Higher-Education Perspectives	Lintang Matahari Hasani, Hafizh Rafizal Adnan,Dana Indra Sensuse, Kautsarina,Ryan Randy Suryono	2020	International Conference On Computer And Informatics Engineering

TITLE OF THE PAPER	AUTHORS	PUBLICATION YEAR	SOURCE TITLE
Understanding Academics' Adoption Of Learning Technologies: A Systematic Review	Qian Liu,Susan Geertshuis, Rebecca Grainger	2020	Computers & Education 151
Multiple Perspective Of Cloud Computing Adoption Determinants In Higher Education A Systematic Review	Mohammed Banu Ali	2019	International Journal Of Cloud Applications And Computing
Investigating Students' Acceptance Of E-Learning System In Higher Educational Environments In The UAE: Applying The Extended Technology Acceptance Model (TAM)	Said Abdelrahim Said Salloum	2018	Doctoral Dissertation
Factors Affecting The Adoption Of E-Learning Technologies Among Higher Education Students In Nigeria: A Structural Equation Modelling Approach	M Nasiru Yakubu , Salihu Ibrahim Dasuki	2018	Information Development
Analysis Of The Factors For The Successful E-Learning Services Adoption From Education Providers' And Students' Perspectives: A Case Study Of Private Universities In Northern Iraq	Mohammed Simko Abdullah , Mehmet Toycan	2017	EURASIA Journal Of Mathematics, Science And Technology Education
Determinants Of E-Learning Adoption In Universities: Evidence From A Developing Country	Eric Ansong, Sheena Lovia Boateng, And Richard Boateng	2017	Journal Of Educational Technology Systems

TITLE OF THE PAPER	AUTHORS	PUBLICATION YEAR	SOURCE TITLE
A Decade Of Lms Acceptance And Adoption Research In Sub-Sahara African Higher Education: A Systematic Review Of Models, Methodologies, Milestones And Main Challenges	Brandford Bervell , Irfan Naufal Umar	2017	EURASIA Journal Of Mathematics, Science And Technology Education
A Review On Cloud Computing Acceptance Factors	Mohd Talmizie Amrona,Roslina Ibrahima, Suriayati Chuprata	2017	Procedia Computer Science
A Systematic Review On Mobile Learning In Higher Education: The African Perspective	Rogers Kaliisa, Michelle Picard	2017	The Turkish Online Journal Of Educational Technology
Factors Affecting E-Learning Adoption In Developing Countries—Empirical Evidence From Pakistan's Higher Education Sector	Faria Kanwal And Mariam Rehman	2017	IEEE Access
Expanding The Technology Acceptance Model (TAM) To Examine Faculty Use Of Learning Management Systems (Lmss) In Higher Education Institutions	Nafsaniath Fathema, David Shannon, Margaret Ross	2015	MERLOT Journal Of Online Learning And Teaching
Exploring The Factors That Affect Student Satisfaction Through Using E-Learning In Malaysian Higher Education Institutions	Waleed Mugahed Al-Rahmi, Mohd Shahizan Othman, Lizawati Mi Yusuf	2015	Mediterranean Journal Of Social Sciences

TITLE OF THE PAPER	AUTHORS	PUBLICATION YEAR	SOURCE TITLE
Barriers And Enablers To Adoption And Diffusion Of Elearning	Gurmak Singh , Glenn Hardaker	2014	Campus-Wide Information Systems
Exploring Factors That Influence Adoption Of E-Learning Within Higher Education	Emma King , Russell Boyatt	2014	British Journal Of Educational Technology

Table 2: Journal Wise Categorization

Journal Name	Frequency
Education And Information Technologies	3
E-Learning And Digital Media	1
International Journal Of Educational Technology In Higher Education	1
Technology, Knowledge And Learning	1
Hindawi Education Research International	1
International Journal Of Interactive Mobile Technologies	1
Multidisciplinary Journal For Education ,Social And Technological Sciences	1
Interactive Learning Environments	1
International Conference On Computer And Informatics Engineering	1
Computers & Education	1
International Journal Of Cloud Applications And Computing	1
Information Development	1
EURASIA Journal Of Mathematics, Science And Technology Education	2
Journal Of Educational Technology Systems	1
Procedia Computer Science	1
The Turkish Online Journal Of Educational Technology	1
IEEE Access	1
MERLOT Journal Of Online Learning And Teaching	1
Mediterranean Journal Of Social Sciences	1
Campus-Wide Information Systems	1
British Journal Of Educational Technology	1

Discussion

This section summarizes and discusses the results and findings of the related articles that we have included in this study. One of the important research questions by this study was to evaluate the research types, methods, countries of research and the theoretical frameworks commonly used in online learning studies within higher education institutions. The main aim of the research was to find out various factors which are behind the readiness to adopt e-learning in higher education institutions. 17 papers included in this study are relevant for this research which contributed by different authors in identifying these factors.

The most used behavioral theories that were used in the previous studies which evaluate the users willingness to adopt any technology, innovation are Theory of planned behavior(TPB), Theory of reasoned action(TRA)(Gupta et al., 2021) .For explaining various behaviours of acceptance with regards to a technology or an innovation, several theories have been emerged and several models were also proposed. The theory of reasoned action (TRA) (Fishbein and Ajzen, 1975), planned behavioral theory (Ajzen 1991), social cognitive theory (Bandura 1977), Diffusion innovation theory(Everett M. Rogers 1995), the Technologies Acceptance Model (TAM) (Davis 1986), the unified theory of acceptance of the use of technologies (UTAUT) (Venkatesh et al. 2003) are some of the used models and theories.

The most used model in the previous studies was Technology Acceptance Model (TAM), Extended Technology Acceptance Model (ETAM).

If we talk about the factors, while studying the previous papers related to online learning, training is one of the major factors which becomes the reason for adopting an innovation or technology in higher education (Aldunate and Nussbaum 2013). To adopt the technology, firstly it is important to purchase the required equipment. To make use of those equipment it is necessary to train the students as well as faculties for a successful adoption of technology (Budin, 1999). Kopcha (2012), in his study, reviewed while studying the factors of e-learning adoption that a teacher's belief is also one of the major factors which influences online teaching adoption. (Ben Romdhane 2013) also explained "satisfaction" as another important factor for accepting e-learning. But satisfaction can be there only after accepting the technology. For being satisfied (Ben Romdhane and Emna, 2013); (Sun et al., 2008) identified interfaces of the technology and the content included were found to be the most used elements for accepting an e-learning technology. This study also reviewed various advantages of using e-learning resources. According to Smedley (2010), e-learning provides the students as well as the users the convenience of time and place. They can learn anywhere , at any time. Another benefit is that e-learning provides self-pacing, which means users can learn at their own pace accordingly (Codone, 2001).

Despite having so many advantages or factors for the acceptance of e-learning, there are various barriers in the path of its acceptance according to various studies. (Assareh and Hosseini Bidokht, 2011) have identified four factors that lead to hurdles for accepting e-learning which are students, faculties, curriculum and the institution. In case of faculties, it can be lack of knowledge about how to use e-learning tools or devices, lack of interest, confidence deficiency can be the reasons for rejection of technology (Bhuasiri et al. 2012). (Ireson et al., 2015) explained resistance to change as one of the major factors for non-acceptance of an innovation or technology. He identified that age can be the reason for their behaviour of not to get changed with time (Ireson et al., 2015). Collins et al. (1997), explained that users spend their time using various social sites for no reasons, this may lead to waste of time and money.

Conclusion

The number of papers based on e-learning adoption are growing rapidly, even though there are selected studies which provide evidence of its effectiveness. Most of the studies have been done on students' perspective related to this study's theme. Less studies have been conducted on faculty perspective. This will be an opportunity for the future scholars to find out the factors and hindrances that are responsible for the adoption of e-learning in higher education institutions. Previous studies have found factors which are responsible for students' readiness for e-learning adoption in higher education institutions such as students' comfort with e-learning, competence of students, uncertainty avoidance , various technological and financial conditions. Therefore, to ensure effective e-learning systems faculties are suggested to encourage the students, to introduce the students with the e-learning resources as soon as possible, to get them engaged with the different applications of e-learning with its usage and applicability, to provide financial assistance to the students if needed, to give the students as well as the instructors clear instructions about the rules and regulation required for effective teaching and learning.

Implications

Educational scenario has completely shifted from traditional teaching methods to online learning tools, distant learning methods. Findings of this study will definitely help the higher education institutions in adopting the educational change for betterment of teaching-learning process. Various suggestions have been provided in the above section which will assist the students as well as the teachers in adopting the new innovations and technologies in education. Higher institutions can study the factors from this paper and it will help them in identifying which factors are helpful in acceptance of e-learning technology and which factors act as barriers which lead to rejection. In order to give a solution to the e-learning problems, it is suggested that online faculties should adopt harmonized learning processes at the University and also to plan student centred curriculum so that they can learn in a better way and faculties will be able to transfer knowledge in a more comprehensive manner.

Limitation

The following are the limitations of the study: Firstly, most of the reviewed studies have been done in urban areas and rural areas were not in the main focus. Educational scenarios can be different according to the culture and region, these theories and models may not work under those conditions. So, a different methodology may be needed to conduct research in rural educational institutions. So the results of this study cannot be generalized. Secondly, barriers to adopt technology have been discussed in the study but solutions to them have not been provided in this study. They are yet to be researched further.

Future Research Directions

Previous studies were mostly limited to the students' perception about the adoption of e-learning. Future research can be done to observe the behavior of faculty to adopt the e-learning system in the classrooms. Taking faculties as an organizational unit for the study will make the research more holistic. Future research can also be done to analyze the problems and issues faced by the users in remote areas who do not get proper access to the internet. Also, they can provide suggestions and recommendations to remove the constraints which will make the teaching-learning process more effective.

Research can also be explored as a comparative study between faculties and students' adoption of e-learning.

Also, most of the studies have been done in urban areas' private higher educational institutions but rural areas were ignored. Due to differences in modes of education, culture, region, individual differences, adequate resources to use e-learning results cannot be generalized. Future research can be done in rural areas' educational institutions as well.

References

Abbad, M. M., Morris, D., & De Nahlik, C. (2009). Looking under the bonnet: Factors affecting student adoption of e-learning systems in Jordan. *International Review of Research in Open and Distributed Learning, 10*(2).

Abdullah, M. S., & Toycan, M. (2017). Analysis of the factors for the successful e-learning services adoption from education providers' and students' perspectives: A case study of private universities in Northern Iraq. *Eurasia Journal of Mathematics, Science and Technology Education, 14*(3), 1097-1109.

Ajzen, I. (1991). The theory of planned behavior. *Organizational behavior and human decision processes, 50*(2), 179-211.

Aldunate, R., & Nussbaum, M. (2013). Teacher adoption of technology. *Computers in Human Behavior, 29*(3), 519-524.

Ali, M. B. (2019). Multiple perspective of cloud computing adoption determinants in higher education a systematic review. *International Journal of Cloud Applications and Computing (IJCAC), 9*(3), 89-109.

Al-Rahmi, W. M., Othman, M. S., & Yusuf, L. M. (2015). Exploring the factors that affect student satisfaction through using e-learning in Malaysian higher education institutions. *Mediterranean Journal of Social Sciences, 6*(4), 299.

Amron, M. T., Ibrahim, R., & Chuprat, S. (2017). A review on cloud computing acceptance factors. *Procedia Computer Science, 124*, 639-646.

Ansong, E., Lovia Boateng, S., & Boateng, R. (2017). Determinants of e-learning adoption in universities: Evidence from a developing country. *Journal of Educational Technology Systems, 46*(1), 30-60.

Anthony, B., Kamaludin, A., Romli, A., Raffei, A. F. M., Phon, D. N. A., Abdullah, A., & Ming, G. L. (2020). Blended learning adoption and implementation in higher education: A theoretical and systematic review. *Technology, Knowledge and Learning*, 1-48.

Assareh, A., & Bidokht, M. H. (2011). Barriers to e-teaching and e-learning. *Procedia Computer Science, 3*, 791-795.

Awan, R. K., Afshan, G., & Memon, A. B. (2021). Adoption of E-Learning at Higher Education Institutions: A Systematic Literature Review.

Multidisciplinary Journal for Education, Social and Technological Sciences, *8*(2), 74-91.

Bandura, A. (1977). Self-efficacy: toward a unifying theory of behavioral change. *Psychological review*, *84*(2), 191.

Basilotta-Gómez-Pablos, V., Matarranz, M., Casado-Aranda, L. A., & Otto, A. (2022). Teachers' digital competencies in higher education: a systematic literature review. *International Journal of Educational Technology in Higher Education*, *19*(1), 1-16.

Ben Romdhane, E. (2013). La question de l'acceptation des outils de e-learning par les apprenants: quels dimensions et déterminants en milieu universitaire tunisien?. *Revue internationale des technologies en pédagogie universitaire/International Journal of Technologies in Higher Education*, *10*(1), 46-57.

Bervell, B., & Umar, I. N. (2017). A decade of LMS acceptance and adoption research in Sub-Sahara African higher education: A systematic review of models, methodologies, milestones and main challenges. *Eurasia Journal of Mathematics, Science and Technology Education*, *13*(11), 7269-7286.

Bhuasiri, W., Xaymoungkhoun, O., Zo, H., Rho, J. J., & Ciganek, A. P. (2012). Critical success factors for e-learning in developing countries: A comparative analysis between ICT experts and faculty. *Computers & Education*, *58*(2), 843-855.

Budin, H. (1999). The computer enters the classroom. *Teachers college record*, *100*(3), 656-669.

Castro, M. D. B., & Tumibay, G. M. (2021). A literature review: efficacy of online learning courses for higher education institution using meta-analysis. *Education and Information Technologies*, *26*(2), 1367-1385.

Codone, S. (2001). An E-Learning Primer, Raytheon Interactive Pensacola, Florida.

Collins, J., Hammond, M., & Wellington, J. (2002). *Teaching and learning with multimedia*. Routledge.

Davis, F. D. (1985). *A technology acceptance model for empirically testing new end-user information systems: Theory and results* (Doctoral dissertation, Massachusetts Institute of Technology).

Fathema, N., Shannon, D., & Ross, M. (2015). Expanding the Technology Acceptance Model (TAM) to examine faculty use of Learning Management Systems (LMSs) in higher education institutions. *Journal of Online Learning & Teaching*, *11*(2).

Fishbem, M., & Ajzen, I. (1975). Belief Attitude, Intention and Behavior An Introduction to Theory and Research Addison Wesley. *Reading, Massachusetts.*

Gama, L. C., Chipeta, G. T., & Chawinga, W. D. (2022). Electronic learning benefits and challenges in Malawi's higher education: A literature review. *Education and Information Technologies*, 1-18.

Grandon, E. E., Alshare, K., & Kwun, O. (2005). Factors influencing student intention to adopt online classes: A cross-cultural study. *Journal of Computing Sciences in Colleges, 20*(4), 46-56.

Gupta, Y., Khan, F. M., & Agarwal, S. (2021). Exploring Factors Influencing Mobile Learning in Higher Education-A Systematic Review. *International Journal of Interactive Mobile Technologies, 15*(12).

Hasani, L. M., Adnan, H. R., Sensuse, D. I., & Suryono, R. R. (2020, September). Factors affecting student's perceived readiness on abrupt distance learning adoption: Indonesian higher-education perspectives. In *2020 3rd International Conference on Computer and Informatics Engineering (IC2IE)* (pp. 286-292). IEEE.

Kaliisa, R., & Picard, M. (2017). A systematic review on mobile learning in higher education: The African perspective. *TOJET: The Turkish Online Journal of Educational Technology, 16*(1).

Kanwal, F., & Rehman, M. (2017). Factors affecting e-learning adoption in developing countries–empirical evidence from Pakistan's higher education sector. *Ieee Access, 5*, 10968-10978.

Khan, M. A., & Khojah, M. (2022). Artificial intelligence and big data: The advent of new pedagogy in the adaptive e-learning system in the higher educational institutions of Saudi Arabia. *Education Research International, 2022.*

King, E., & Boyatt, R. (2015). Exploring factors that influence adoption of e-learning within higher education. *British Journal of Educational Technology, 46*(6), 1272-1280.

Kisanga, D., & Ireson, G. (2015). Barriers and strategies on adoption of e-learning in Tanzanian higher learning institutions: Lessons for adopters. *International Journal of Education and Development using ICT, 11*(2).

Kopcha, T. J. (2012). Teachers' perceptions of the barriers to technology integration and practices with technology under situated professional development. *Computers & Education, 59*(4), 1109-1121.

Liu, Q., Geertshuis, S., & Grainger, R. (2020). Understanding academics' adoption of learning technologies: A systematic review. *Computers & Education, 151*, 103857.

Memon, A. B., & Meyer, K. (2017). Why we need dedicated web-based collaboration platforms for inter-organizational connectivity? A research synthesis. *International Journal of Information Technology and Computer Science, 9*(11), 1-11.

Ndibalema, P. (2022). Constraints of transition to online distance learning in Higher Education Institutions during COVID-19 in developing countries: A systematic review. *E-Learning and Digital Media,* 20427530221107510.

Prensky, M. (2001). Digital Natives, Digital Immigrants. OntheHorizon, 9 (5), 1-6.

Robey, D., & Markus, M. L. (1998). Beyond rigor and relevance: Producing consumable research about information systems. *Information Resources Management Journal (IRMJ)*, *11*(1), 7-16.

Rogers, E. M. (1995). Diffusion of Innovations: modifications of a model for telecommunications. In *Die diffusion von innovationen in der telekommunikation* (pp. 25-38). Springer, Berlin, Heidelberg.

Salloum, S. A. S. (2018). *Investigating students' acceptance of e-learning system in higher educational environments in the UAE: Applying the extended technology acceptance model (TAM)* (Doctoral dissertation, The British University in Dubai).

Sayaf, A. M., Alamri, M. M., Alqahtani, M. A., & Al-Rahmi, W. M. (2021). Information and communications technology used in higher education: An empirical study on digital learning as sustainability. *Sustainability*, *13*(13), 7074.

Singh, G., & Hardaker, G. (2014). Barriers and enablers to adoption and diffusion of eLearning: A systematic review of the literature–a need for an integrative approach. *Education+ Training.*

Smedley, J. (2010). Modelling the impact of knowledge management using technology. *OR insight*, *23*(4), 233-250.

Sun, P. C., Tsai, R. J., Finger, G., Chen, Y. Y., & Yeh, D. (2008). What drives a successful e-Learning? An empirical investigation of the critical factors influencing learner satisfaction. *Computers & education*, *50*(4), 1183-1202.

Vanitha, P. S., & Alathur, S. (2021). Factors influencing E-learning adoption in India: Learners' perspective. *Education and Information Technologies*, *26*(5), 5199-5236.

Venkatesh, V., Morris, M. G., Davis, G. B., & Davis, F. D. (2003). User acceptance of information technology: Toward a unified view. *MIS quarterly*, 425-478.

Yakubu, M. N., & Dasuki, S. I. (2019). Factors affecting the adoption of e-learning technologies among higher education students in Nigeria: A structural equation modelling approach. *Information Development*, *35*(3), 492-502.

Zhang, Z., Cao, T., Shu, J., & Liu, H. (2020). Identifying key factors affecting college students' adoption of the e-learning system in mandatory blended learning environments. *Interactive Learning Environments*, 1-14.

❑❑❑

16

Internationalization of Higher Education, Regional Cooperation and Digital Integration: A Perspective

Dr. S. K. Bose

Associate Professor, School of Law, Manav Rachna University

Abstract

Time is among the most important factors in schooling. The goal of education cannot be claimed to have been achieved if sufficient change is not brought about in the system within a reasonable amount of time or if it takes too long to determine it. According to a saying, stasis leads to death and change is the law of nature. If we examine the elements pertaining to the regulation of digital education, or the examples pertaining to this notion, we see that this is one of the sectors that is most in development. Internationalization and digital education are given a lot of attention in the National Education Policy 2020. Undoubtedly, a digital education empowers individuals to engage in social life in the future on their own terms, ensuring the competitiveness of the businesses. The difference between what we learn and how we learn must be drastically discovered in order to keep up with the digital transformation. As a result, in addition to reading, writing, and math, digital abilities are now crucial to success in general.

Introduction

One of the most crucial aspects of education is that of time. If education does not bring significant change to the system within the right time frame, or if too much time is taken in determining it, the very purpose of it cannot be said to have been served. It is said that change is the law of nature and stagnation is death. If we look at the aspects relating to the regulation of digital education, or at the instances relating to this concept, we find that this is one of the most evolving areas. The National Education Policy 2020 lays much emphasis on digital education and internationalization. Certainly, a digital education enables people to participate in social life in a self-determined manner in the future which ensures the companies to remain competitive. Going with the flow of digital transformation, one must radically find the difference between what we learn and how we learn. Consequently, notwithstanding reading, writing and calculations, digital skills have become extremely critical to overall success. But all these must adhere to the time period. In the era of fast moving technology, lightning-fast data assessment facilities make it possible to cater to the individual needs of the students. There are, of course, soft wares which can keep the track of progress and accordingly monitor the developments. This also helps us customize the courses and identify the gaps which need to be plugged in. While internationalization of education is taking a huge leap as seen worldwide, digital education needs more time to become a reality. Manpower, that too resourceful manpower, is one of the major pre-requisites of integration of digital education. Manpower, which I am referring to, means trained teaching faculty who can use digital media to impart knowledge to the students and effectively transfer relevant information to them. There are many issues and challenges with regard to internationalization and digitalization of education. However, this chapter depicts very limited points which are of greater significance.

Challenging Dynamics of Internationalization

Internationalization of higher education has always been a subject of animated debate among competing minds and advanced nations. This brings the realists to ponder over various aspects relating to the core of the subject so as to rediscover more though there is a well established set of ideas for understanding the present system of education. For further guiding policy implementation, it follows the same set of ideas. The dynamics and nuances of internationalization of higher education may change considerably but the logic remains as it is. Considering this angle, one should certainly care about power transitions. Of course, bringing sea changes in the system is always a difficult proposition.

Emerging education entities invariably enquire about various hurdles faced by the established players and challenge their status quo. They strive to bring drastic changes in the dominant norms of the system in order to give momentum to their own interests and practices. This is, by no means an easy task to accomplish. There is always an uncontrollable temptation for the established actors to use their powers and strategies to block the emergence of rising nations. Naturally, there arise conflicts and tensions among the nations, and the experts are divided in their opinions with regard to these conflicts that broadly revolve around the power shifting or power transition. Furthermore, the theorists differ as to whether conflict derives more from the actions of revisionist powers that are pioneering to rebuild the rules of international affairs- that includes socio, economic and educational cultural ties, or from the status quo powers which have growing apprehensions to preserve their supremacy.

However, experts from the realist camp have a consensus and believe that rising powers or the revisionists should be appreciated and acknowledged for their positive impact on the global platform and these transitions are not only terrifying but disquieting also. For some, revised systems always matter as their growing stature influences the balance of international affairs as they bring disruptions. Needless to mention, there have been unending discussions and debates with respect to power transitions and conspicuous changes in the system that give rise to conflicts and controversies. Some believe this is natural while others don't. Nonetheless, there is a broad consensus as far as power shifts are concerned which have always been linked to conflicts. This too will continue. Subsequently, in the matters of international relationship behavior of nations, particularly emerging nations, their status as well as recognition is extremely critical. Advocates from neo-realist schools reckon that existing order too is a vital factor in international affairs including internationalization of education. Despite acknowledging the notion of rising education status of certain countries as eminent, it is beyond one's comprehension as to how one can objectively major the sources of their disillusionment.

Regional Cooperation and Interdependence

In the realm of internationalization of higher education, one must not forget that regional cooperation plays a pivotal role as it aims for an overall growth of societal integration and the process of socio-economic interaction for creating a vibrant ambience. This may produce interdependence, creating an ecosystem of common interests and identity. Of late, this regional interdependence has taken different forms of interaction as nations promote regional cohesion in the field of education in diverse ways as nation-led actions, in some situations may have been responsible for a growing actual interaction while in other situations, these relations meant for growth and development may not serve the real purpose rather end up being a mere market led integration.

According to Edward Best & Thomas Christiansen, "there is a distinction between cooperation and integration, considering the different kinds of arrangements that may be agreed among nations. There are various forms of regional cooperation. Functional cooperation is the one that refers to limited arrangements that are agreed among nations in order to work together in particular areas, for example in transport, education, or health. Economic cooperation refers to agreements that foresee some degree of commercial preferences. Political cooperation involves mutual support and commitment regarding the implementation of certain values and practices within the countries." This also means, in the sphere of internationalization of higher education, nation-states, their government agencies exchange thoughts, share ideas, formally or informally inform and consult each other for adoption of common practices, policies while deliberating and selecting certain parameters for the overall dissemination of studies and researches. This also includes taking positions in international platforms in addition to implementing joint actions elsewhere. In connection with different types of cooperation, beyond usual obligations under the international law, there exist neither any other connections nor any significant consequential derivations regarding international status of participating nations. But there exist several other obstacles which do not help the cause. This brings me to discuss further about formal regional integration which refers to a process that helps nation states not only make collective efforts for eliminating hindrances in order to create a regional space but also go beyond for ease of business subject to the commonly accepted rules which also means the nations not only work in cohesion to practice measures of positive integration but also voice universally in unison in the making of a new world of higher education. Consequently, cooperation and integration may not be thought of without the concept of fundamental choices. However, with all seriousness they should be treated differently and carefully. All regional organizations including the EU have a mixture of both cooperation and integration. Regional mechanisms often focus on guaranteeing peace and security response to regionalization whereas the management of internationalization-the interrelationship between regional arrangements and the rest of the world which focus on promoting internal and international dynamics leading to multilateralism.

Systemic Cooperation with Reference to India and European Nations

The approaches to growing competing-interests and global order can bring abundant change to the system. Though they are very productive and influential, they do not always subscribe to positive environmental breakthroughs. One can see them as potential pathways that might lead to the rising power competition. Interestingly, there is no substantive evidence to explain as to how an international system might make a strong move from the gross transmission of power to a scenario of multi-polarity, to a system in which the education policies of the established nations are influenced by common-interest-politics and rationale. Such systems do not at once emerge from anywhere but need systemic cooperation and a high degree of integration. Here come to my mind broadly two ways in which a global practice in higher education might come into existence. As Edward Best & Thomas Christiansen suggest, "one of the ways is through consensus and standing together on almost equal terms of a series of regionally-based systems, whether composed of nations or other groups. The other is by the global dominance of what was originally a regional system." Let's look at the following example: recently India and Germany launched an Indo-German Partnership for Green and Sustainable Development. Under this, Germany has committed to provide additional development assistance of 10 billion dollars till 2030 to support India's green growth plans. This assistance will be provided to India under the Joint Declaration of Intent on Green and Sustainable Development Partnership signed by both India and Germany in Berlin during Mr. Modi's visit to Europe. Both countries acknowledged their joint responsibility for the protection of the planet and for shared, sustainable and inclusive growth, leaving no one behind. The financial and technical cooperation and other assistance to India by Germany will help in achieving the goals related to climate action and sustainable development space, further promoting German-Indian Research and Development (R&D), and encouraging private investment. Further, both sides will work towards identifying deliverables in the priority areas of sustainable urban development, energy transition and environmental protection etc.

Both India and Germany welcomed the finalization of negotiations on a bilateral agreement for a Comprehensive Migration and Mobility Partnership (CMMP) and agreed to take action to swiftly sign the agreement and bring it into force. The importance of this agreement is: facilitating two way mobility of students, professionals and researchers also addressing the challenges of illegal migration.

The Prime Minister of India deliberated that many countries had benefited from India's skilled workers and professionals and the migration and mobility agreement with Germany will facilitate movement between the two sides at a time when India is witnessing the fastest growth in the post Covid19 era as compared to the other developing economies. This shows that systemic cooperation and regional integration are a key to sustainable growth of internationalization of education.

Conclusion

I have attempted in this chapter to characterize the notion of internationalization of education, briefly touching upon digital education and regional integration along with some other aspects of the notion, which is relevant to the overall sustainability of modern education as envisaged by NEP 2020. In doing so, various nuances and dynamics of education have also been discussed. Once we understand this concept in a holistic manner, it would be evident that if education were to develop and sustain further, a qualitative change needs to be made in the implementation and application of those policies and decisions. As discussed Internationalization of higher education has always been debated in advanced nations among intellectuals. The realists think over various aspects relating to this subject so as to rediscover more ideas for understanding the present system of education. As far as power shifting or power transition is concerned, the theorists differ on the matter of conflict which derives more from the actions of revisionist powers those are pioneering to rebuild the rules of international affairs- that includes socio, economic and educational cultural ties, or from the status quo powers which have growing apprehensions to preserve their supremacy. Various forms of regional cooperation have also been touched upon. Functional cooperation which refers to limited arrangements that are agreed among nations in order to work together in particular areas has been discussed. Economic cooperation, Political cooperation which involves mutual support and commitment regarding the implementation of certain values and practices within the countries are also deliberated. This also means, in the sphere of internationalization of higher education, nation-states, their government agencies exchange thoughts, share ideas, formally or informally inform and consult each other for adoption of common practices, policies while deliberating and selecting certain parameters for the overall dissemination of studies and researches. Global practices and importance of research and development in the realms of digital and internationalization of education have also been emphasized.

References

A Comprehensive Training Policy for technical teachers, 2018, ALL INDIA COUNCIL FOR TECHNICAL EDUCATION, Nelson Mandela Marg, New Delhi.

Tahira Jibeen and MA Khan, Internationalisation of Higher Education: Potential Benefits and Costs, International Journal of Evaluation and Research in Education, Vol. 4, No. 4, December 2015.

Whitaker, AM.,"The Internationalization ofhigher education: A US perspective", 2004. Retrieved from http://scholar.lib.vt.edu/theses/available/etd-06202004 192329/ unrestricted/WhitakerMP.pdf.

Polan-Egron, E.,"Higher education internationalization: Seeking a newbalance of values",2012. Retrieved fromhttps://www.nafsa.org/Explore_International_ Education/Trends//TI/Higher_Education_internationalization_Seeking_a_New_Balance_of_Values/.

Cohen, A., Yemini, M., Sadeh, E., "Web-based analysis of Internationalization in Israeli teaching colleges", Journal of Studies in Internationalization Education, 2013.Doi: 10.1177/1028315313479131.

Wildavsky, B.,"The great brain race: How global university are reshaping the world", Princeton University Press, 2010.

AACSB,"The globalization of management education", Emerald Group, Publishing Bingley: UK, 2011.

Knight, J.,"Internationalization:Elements and checkpoints", Research Monograph, no. 7, Ottawa, Canada.Canadian Bureau for International Education, 1994.

Knight, J., Hans, de Wit.,"Internationalization of higher education in Asia Pacific countries", Amsterdam: European Association for InternationalizationEducation Publications, 1997.

Altbach, PG., Knight, J., "The Internationalization of higher education: Motivations and realities", Boston CollegeCenter for International Higher Education, 2007. http://www.bc.edu/bc_org/avp/soe/cihe/pga/pdf/AltbachKnight_2007.pdf.

Kreber, C.,"Different Perspectives on Internationalization in Higher Education",New Directions for Teaching and Learning,vol. 118, pp. 1-14, 2009.

Edmonds, LJ.,"What internationalization should really be about?", 2012. Retrieved from http://www.universityaffairs.ca/what-internationalization-should really-be-about. aspx.

□□□

17

Digital Education: Significance and Challenges in the 21st Century

Dr. Parul Jhajharia
Professor & Dean Faculty of Management & Commerce Manav Rachna University

Dr. Rajnish Ratna
Associate Professor (HR & BRM) Gedu College of Business Studies Royal University of Bhutan

Background and Introduction

There is no doubt that education is changing rapidly. Gone are the days when a degree was the only route to success. Now, there are dozens of options for getting an education that can benefit you in various ways. From online degrees to programmable learning tools, there is something for everyone. Moreover, with so many options, it can be hard to know where to start. This chapter is aimed at the significance and challenges of digital education and some of its key points.

It is safe to say that India's educational system has been impacted by the recent COVID-19 outbreak and subsequent lockdowns. Students could not keep up with their schoolwork while everyone was trapped inside their homes for months. The celebrations they had planned with joy turned out to be chaotic. Realizing it had deprived them of their education, the government sought a solution and settled on providing access to educational materials online.

There has been a dramatic shift in the face of India's educational landscape as more and more students take classes online. The education system has seen significant changes since the New

Education Policy (NEP) came into effect in 2020, including online education; yet, several challenges come with a new set of laws -

- UNESCO reports that the shutdown of schools and colleges has impacted 1.37 billion students in 138 countries worldwide due to the COVID-19 pandemic. In addition, about 60.2 million professors and professionals no longer have to teach students physically.
- Online courses are the best and only choice now that the world is under lockdown. Faculty members throughout the country are signing up for video conferencing services like Zoom, Skype, Google Classroom, and Meet to better communicate with their students.

Although this new medium offers the promise of any time, everywhere access to information across all digital platforms, it proves to be a challenge for both the administration and the students. Traditional print materials like books and notebooks coexist with newer forms of digital media like eBooks and pdfs to create a hybrid learning environment.

The Pandemic of COVID-19 and the Value of Distance Learning

High rates of development and acceptance in education technology were already present before COVID-19, with worldwide edtech investments hitting US$18.66 billion in 2019 and the whole industry for online education estimated to reach $350 Billion by 2025. There has been a massive increase in online educational resources such as language applications, virtual tutoring, video conferencing, and online learning software since COVID-19.

Since lockdown was implemented in March 2020, owing to the nationwide spread of coronavirus, students' lives have revolved almost entirely around their attendance in online classrooms. This "*new normal*" is a radically revised idea of schooling, with online education at its centre.

Students have the flexibility to take courses from wherever they happen to be, and institutions get access to a far larger pool of potential students without having to worry about transportation to and from campus. In addition, students may have their questions answered without leaving the safety of their homes by participating in online discussion groups or live chats.

Teaching and learning have benefited from the increased technology use of online education. For example, many educators turn to technology-based programs and apps for smartphones and computers when schools are on lockdown. Likewise, the school's faculty and student body adapted quickly to using laptops and mobile devices.

However, even though India's biological and cultural diversity is well-known, few people realize that the country also has a significant economic difference. Many people still lack easy access to online education because of obstacles such as a lack of reliable electricity, poor or expensive internet connections, or an inability to buy the gadgets required to participate. A strong and thorough approach is urgently needed to address these issues at the institutional level. Online schooling has become a viable option during the pandemic's mayhem. It is no longer a luxury but a need. As a result, we are at a pivotal juncture in the effort to improve the quality of online teaching and learning.

EdTech startups are flooding the online education industry, intending to upset the status quo. In the face of this difficulty, several are seizing the moment by offering kids free access to online education.

Significance of Online Education

With the introduction of online education, you will be able to avoid the costs of travel, hotel, and boarding, which will translate to lower online course prices for everyone.

- Digital classrooms, in contrast to traditional ones, are unrestricted in terms of space and equipment.
- Unlike traditional classrooms, where teachers are only allowed to reach out to students in their direct proximity, those running online courses are not restricted to serving just their own country's students. Regarding faculty, we will not be limited in any way and can find and employ the most qualified individuals from anywhere in the world.
- Environmentally beneficial; switching to digital is good for the planet. In addition, since the paper is made from trees, reducing textbook printing would significantly impact deforestation rates.
- Time saver: Unlike the traditional teaching methods, which need years of study and months of waiting for the results of the final exams, the digital implementation of the courses allows us to access the material immediately, administer

exams while comfortably seated at home, and get the results quickly (or within a few days) on our smartphones and computers. This reduces stress and saves a great deal of time.

Challenges Facing the Adoption of Online Education

However, there are several obstacles to overcome when integrating online learning into the traditional educational system.

- Many people in rural areas of India, for instance, do not know how to turn on a computer, much alone use one. If we are serious about bringing online education to all parts of the nation, everyone must have at least a familiarity with computers.
- There is a severe problem with the price. It might be difficult for those from lower socioeconomic backgrounds to purchase a laptop or computer, such as farmers, maids, home employees, and sweepers.
- In addition, this presents a significant barrier for classroom teachers. Being effective in online sessions is not as crucial as being a classroom instructor.
- Learning by doing is not an option in virtual classrooms. In topics like physics and the practical arts, students often require hands-on experience in addition to theoretical explanations from their teachers before they can fully absorb the material.
- The number of questions on an online test is restricted in India due to a lack of resources, and the same is valid for the country's ability to create new questions.
- It is not feasible to have live sessions and broadcast them to kids in rural India due to the lack of reliable internet service in such areas. Even though residents in remote regions are learning about and making efforts to access modern technology like smartphones, computers, and tablets, a lack of internet network supply remains a significant barrier to the widespread adoption of these tools.

Major Issues with Classes Online

Socioeconomic Division

The significant socioeconomic gap in India, which includes the split between high, low, and middle classes, directly results from

India's cultural and religious diversity. Only around a third of people worldwide have access to online courses.

Lack of internet access, inadequate electricity, and a lack of available funds to purchase necessary equipment are all pressing issues in rural communities. In addition, many of their educators cited the inability of students from lower socioeconomic backgrounds to participate in live classes as a source of frustration.

Many schools now offer online courses where students watch recorded lectures on platforms like YouTube and WhatsApp at their leisure. Still, this method is not without problems, such as students failing to grasp the material. Therefore, individuals continue to lack access to information that should be made available.

The Polarization of Genders

Another issue is the gender gap that exists in our society. Out of 733 students surveyed at Bihar's public schools, 28% of females and 36% of boys had cell phones. Most of the time, the adults who had smartphones were men, making them harder for females to get their hands on.

Half of these households could not afford internet access, so their kids were limited to watching educational programming on television. However, it was found that females, unlike guys, spent more time on home tasks, which frequently coincided with the broadcast time.

In many of these situations, parents denied conversing with their daughters and instead volunteered to do the survey themselves. However, if they did manage to communicate with a female kid, it was not easy since they were embarrassed to do so in front of a male family member, highlighting the challenges girls face when using the internet for schoolwork.

Consequences of the Internet

Using the web as a learning resource is difficult. While Google, Firefox, and other browsers provide fast access to a wealth of information, they may be cumbersome to use for online learning. In addition, it is unreasonable to expect educators with a background in traditional institutions like schools and colleges to be proficient in developing and delivering online digital material while ensuring that their students quickly adjust to it.

They find reading signals such as body language and eye contact challenging during an online lecture. The topic of whether or not the pupils are paying attention in class is one of many that emerge.

Can they make sense of what is being said throughout this session? Do you think they are going fast enough? While these are prevalent concerns in more conventional classroom settings, they might be more challenging to address in an online meeting.

Students' Mental Well-being

Young children (ages 5 to 10) have difficulty focusing since their attention span is short; thus, keeping their attention requires much physical activity. However, college students have voiced similar concerns, proving that the problem is not limited to the younger generation.

They seem to place more importance on traditional, hands-on teaching methods than digital ones. However, theoretical instruction in fields like science and technology cannot replace the need for, or the impossibility of, other practical experience gained via laboratory work, thesis projects, and field visits.

Too Much Time Spent in Front of a Screen

Parents of the children who take these sessions also have concerns about their children's exposure to media. Kids' mental health will suffer as they depend more on electronic devices to complete their schoolwork, homework, and other requirements. Their eyes and minds are being taxed. It was limited to 30 minutes for kindergartners and two sessions of 45 minutes for kids in grades one through eight due to the government's relaxation of pandemic standards and the rise of online classrooms.

The New Requirements of Education

This pandemic has thrown a wrench into an education system many felt was on the verge of becoming outdated. The scholar Yuval Noah Harari asserts in his book 21 Lessons for the 21st Century that schools emphasize memorizing facts and formulas rather than developing students' critical thinking and flexibility, two abilities that will be more valuable in the future workplace. The shift to online education raises the question of whether or not it may serve as the impetus for developing a novel and perhaps more efficient approach to teaching. Others, having seen the advantages of e-learning first-hand, want to make it part of their "new normal," while others fear that the rapid pace of the transfer online may have hampered this objective.

COVID-19 Emphasizes the Value of Sharing Information

The growth of e-commerce in the wake of the spread of SARS is a prime illustration of how fast innovation may follow a major global catastrophe. E-learning is one of the few industries where investment has not dried up. However, we do not yet know whether this will be the case after COVID-19. The lessons of this epidemic emphasize the value of sharing information across national boundaries, organizational boundaries, and all sectors of society. All of us are responsible for investigating the possibilities of online learning technology if it can help.

Conclusions

No doubt that education has changed over the years. Gone are the days when you needed to take a whole year off from your existing commitments to pursue higher education. With digital education, you can get started today in just a few clicks! Many institutions have also developed flexible and customizable programs for working professionals who want higher edges but do not have enough time for full-time studies.

Digital education has many benefits and has changed the way people learn in such a short time. There are already changes in the job market due to these new tools. It is becoming increasingly popular for companies to provide their employees with online degrees instead of giving them just a few months' worth of training before placing them in front of a desk.

With all these advancements coming at us so quickly, it is no wonder why so many are shifting towards digital education!

References

Mayes, R., Natividad, G., & Spector, J. M. (2015). Challenges for educational technologists in the 21st century. *Education Sciences, 5*(3), 221-237.

Mishra, P., & Mehta, R. (2017). What we educators get wrong about 21st-century learning: Results of a survey. *Journal of Digital learning in Teacher education, 33*(1), 6-19.

Sukmayadi, V., & Yahya, A. (2020). Indonesian education landscape and the 21st century challenges. *Journal of Social Studies Education Research, 11*(4), 219-234.

Voogt, J., Erstad, O., Dede, C., & Mishra, P. (2013). Challenges to learning and schooling in the digital networked world of the 21st century. *Journal of computer assisted learning, 29*(5), 403-413.

Young, J. A., McLeod, D. A., & Brady, S. R. (2018). The ethics challenge: 21st century social work education, social media, and digital literacies. *The Journal of Social Work Values and Ethics, 15*(1), 13-22.

❑❑❑

SECTION - 6
Solar Energy Sector

18

Role of Artificial Intelligence in Solar Energy Sector: An Appraisal

Pritee Deotale
Self - Practicing advocate

Abstract

Artificial intelligence has brought a new dawn in the domain of science and technology. It has changed the dynamics of every sector keeping no stone unturned. Artificial intelligence has left its mark significantly on the solar energy sector as well. It can easily predict the demand as well as supply of Solar Energy. It is also used for the detection of electricity theft which is a major reason for huge revenue losses for the Indian energy sector. Artificial Intelligence has proved to be a boon for the energy sector across the globe and here this chapter will discuss the significant changes it has brought to the Solar Energy Sector.

Keywords: *Artificial Intelligence, Solar Energy, electricity theft.*

Introduction

Artificial Intelligence has turned out to be a game Changer for the Renewable Energy Sector. Looking at the Sustainable Development Goals (SDG's) also known as Global goals gave the 2030 agenda for Sustainable Development, adopted by the United Nations in 2015. The Paris Agreement i.e. The United Nations Climate Change Conference (COP21) which is a legally binding international treaty on Climate Change 2015, adopted by 196 parties. The COP 21 has come up with a long-term goal for all the parties to it. The International

Solar Alliance 2015 also aims to promote solar energy, 120 nations affirm to participate in this alliance. In India solar energy has the largest contribution in the renewable energy market of India. The National Action Plan of Climate Change (NAPCC) has eight sub-missions and one of them is the National Solar Mission. The National Solar Mission (NSM), launched in 2010 targets 100 GW grid-connected solar plants by 2022. All these treaties talk about the promotion of clean energy, energy efficiency, energy security, energy affordability and reducing carbon footprints. In these initiatives artificial intelligence has proved to be a catalyst in achieving the aim of clean energy by catering important technological solutions.

Challenges Faced by Solar Energy Sector

Solar Energy source comes with various benefits but it also faces certain challenges and those challenges are unpredictable weather conditions. The sudden changes in climate affects the energy flow. This Supply chain of Solar energy is prone to such vulnerabilities. The supply chain needs to be smoothened enough to prevent the unexpected changes. The recent development of solar energy storage is at its nascent stage and it requires thorough testing. It can be seen that the demand for solar energy is increasing thus implementation of artificial intelligence will improve productivity and overcome all the shortfalls.

Role of Artificial Intelligence

The Solar Electric grid is one of the complex machines. It faces many challenges in accommodating the internal variability of solar due to unexpected weather changes. Thus it becomes very necessary to have smart systems. Here's how artificial intelligence improves the reliability of solar energy and modernizes the grid.

- **Capturing Solar Irradiance:** A smart tracking feature with local data insight can provide a tracker control algorithm to determine the installation that would prove best to capture solar radiance. Traditional photovoltaic systems operate on a modular design with pre-established tracking solar irradiance pathways. Artificial intelligence has the potential to change the orientation of panels to capture optimum solar irradiance even when the cloudy skies cause dynamic changes to the environment. This optimized approach can deliver real benefits to solar PV installations.

- **Smart and Centralized Grid:** Most of the devices and grid are interconnected to collect huge amounts of data. The integration of Artificial intelligence can give new insight to the grid operators. It offers flexibility to the energy suppliers to significantly adjust the supply with demand.
- **Grid Management:** A critical aspect of a solar energy system is grid management. Artificial intelligence is playing a pivotal role in this area as well. These technologies use data analytics to predict energy consumption in households. The prediction is based on the specific part of a year and also considers previous years data. This helps power companies stay informed about how much energy will be required in the upcoming days. Based on that, they can manage their grids without any outage. If the consumption is going to be high, they can ramp up energy production. Alternatively, in some parts of the year, when energy consumption is low, they can lower the production to avoid wastage.
- **Improved integration of Microgrids:** The Artificial intelligence powered controlled optimization can solve power quality issues. Intelligent algorithms have the potential to provide real-time control over the dynamic nature of the grid below the substation autonomous controls.
- **Predicting Power Production:** One more benefit is Artificial Intelligence has become very accurate in predicting the power generation from solar photovoltaic installations. Earlier this used to be a hurdle for renewable energy installations. Variable energy production can create challenges for local grid operators attempting to match demand and supply. Companies like Solcast in Australia are taking this AI weather adaptation a step further, using AI forecasting combined with meteorological data to provide system-wide forecasting for solar energy generation. When combined with on-site adaptive Artificial Intelligence forecasting solutions, this can provide an important opportunity to further enhance predictive capabilities around solar energy generation
- **Managing Grid Harmonics:** Artificial Intelligence also helps to manage and mitigate challenges around grid harmonics; in simple terms that is the variable distortions in the frequency of the grid. When these harmonics become misaligned, power system faults can occur. Artificial Intelligence offers a chance to predict, manage, and monitor this to further enhance integration with centralised electricity grids.

- **Optimizing Operational Grids:** Artificial Intelligence technologies also offer a route to enhanced operations and management that can reduce on-going operational costs for solar farms. Smart Artificial Intelligence systems can be taught to adapt and identify failure, providing predictive maintenance solutions that enable operators to undertake cost-saving maintenance before problems emerge. With advanced learning capabilities, Artificial Intelligence systems can also adapt and learn as they gather data through operations. This technology reportedly reduces maintenance analysis and fault identification from potentially a week to just half a day.
- **Expand the Market:** Integration of artificial Intelligence will help the suppliers expand the marketplace by introducing new models and encourage higher participation. Solar Energy data collection is also analyzed by artificial intelligence. Such technologies can also provide insights regarding energy consumption. This helps suppliers to optimize the existing services and launch new service models.
- **Intelligent Energy Storage:** The storage technology is very helpful to solve problems related to the nature of solar energy. There needs to be a proper control over the production, consumption and storage of solar energy. This can only be done when sophisticated software are available. This is where AI plays a vital role, by pairing renewable energy with Artificial Intelligence driven storage we can make a change in paradigm.
- **Maintenance:** No matter how well power grids are managed, there are times when they need maintenance. It is crucial to run the entire system efficiently. By leveraging the power of Artificial Intelligence The specific part of the system that needs maintenance can be easily predicted. When power companies are updated with upcoming maintenance work, they can notify consumers about maintenance in the grid. Scheduled maintenance means consumers can be aware of the forthcoming power cuts. What we witness currently is power cuts without any early announcements.
- **Eliminating RE Barriers and Strengthening Energy Security**: The traditional locations and footprints of solar photovoltaic installations also give rise to several issues which artificial intelligence can help tackle. Cabling is often significant across large solar photovoltaic installations, connecting outer panels with centralized transformer stations. These cables can potentially run up to hundreds of meters

in length, making them vulnerable to problems of electrical arcing which can damage systems. Artificial intelligence technologies can quickly detect and react to arc faults which could cause dangerous electrical damage. Quick-break solutions can trigger to shut down affected elements of the system, providing rapid safety response when needed. These AI-enabled safety features could not only reduce costs by eliminating costly faults, but deliver peace of mind for both rural and urban installation settings. Artificial intelligence is even being used to identify the intriguing challenge of ecology aiming to boost the ecological protections of these systems.

Conclusion

Artificial Intelligence has the potential to reshape the solar energy industry completely. We are already seeing the differences it has made. In the upcoming years, these technologies are going to impact both power companies and consumers. Power companies will have a tool for better forecasting, management of grids and, most importantly, scheduling maintenance. For consumers, the impact will be in the form of interrupted green energy, as well as upfront updates about scheduled maintenance in the grid. In the near future, Artificial Intelligence is expected to benefit the solar energy industry in many more ways.

Suggestions

The technology of Artificial Intelligence has paved a new way for the development of energy generation and the energy market. Bringing technologically advanced programs will significantly help in tapping solar energy and storage. One thing which is nowhere addressed is; the waste created by the solar photovoltaic cells. The waste from solar photovoltaic cells may prove very harmful for the environment as they may have chemical reactions with the elements present in nature if they are left untreated after their use. This will have an adverse effect on the ecology. Thus it needs to be taken care of.

References

United Nations. (2022, July 24). Department of Economics and Social Affairs, *Sustainable Development Goals.* https://sdgs.un.org/goals.

United Nations. (2022, July 24). United Nations Climate Change Conference, *Key Aspects of the Paris Agreement.* https://unfccc.int/process-and-meetings/the-paris-agreement/the-paris-agreement/key-aspects-of-the-paris-agreement.

International Solar Alliance. (2022, July 24). *Paris Declaration on the International Solar Alliance of 30th November 2015.* https://isolaralliance.org/.

Press Information Bureau. (2022, July 24). *National Action Plan on Climate Change.* https://static.pib.gov.in/WriteReadData/specificdocs/documents/2021/dec/doc202112101.pdf

Ashok K. Laha. (2022 July 25). The Role of AI and ML in Solar Energy. *Saur Energy International.*https://www.saurenergy.com/solar-energy-articles/the-role-of-ai-and-ml-in-solar-energy.

Michael Georgiou. (2022 July 25).The Role of AI Technology in Improving the Renewable Energy Sector, *Imaginovation Insider.* https://imaginovation.net/blog/artificial-intelligence-in-renewable-energy/.

Nastel Intelligence from Integration. (2022 July 25). How Artificial Intelligence and Machine Learning are Transforming the Future of Renewable Energy. https://www.nastel.com/how-artificial-intelligence-and-machine-learning-are-transforming-the-future-of-renewable-energy/.

Rohit Trivedi & Shafi Khadem. (2022 July 25). Implementation of artificial intelligence techniques in microgrid control environment: Current progress and future scopes, Elsevier. https://www.sciencedirect.com/science/article/pii/S266654682200009X

Energy Watch. (2022 July 25). 6 Ways of Artificial Intelligence is shaping Our Solar Power Future. *Energy Watch. https://www.energywatch.com.my/blog/2020/11/17/6-ways-artificial-intelligence-is-shaping-our-solar-power-future/.*

Thierry Mortier. (2022 July 26). Why Artificial Intelligence is a game-changer for Renewable Energy, *Renewable Energy Country Attractiveness Index (RECAI).* https://www.ey.com/en_gl/power-utilities/why-artificial-intelligence-is-a-game-changer-for-renewable-energy.

Ashok K. Laha (2022 July 26). The Role of AI and ML in Solar Energy. *Saur Energy International.* https://www.saurenergy.com/solar-energy-articles/the-role-of-ai-and-ml-in-solar-energy.

Radian Belu. (2022 July 26). Artificial Intelligence Techniques for Solar Energy and Photovoltaic Applications. *ResearchGate.* https://www.researchgate.net/publication/286141950_Artificial_intelligence_techniques_for_solar_energy_and_photovoltaic_applications.

Harish Ganjieh. (2022 July 27). How Artificial Intelligence and Machine Learning are Transforming the Future of Renewable Energy. *Forbes Technological Council.* https://www.forbes.com/sites/forbestechcouncil/2021/11/30/how-artificial-intelligence-and-machine-learning-are-transforming-the-future-of-renewable-energy/?sh=ed0ee96541bf.

Ali Q. Al-Shetwi. (2022 July 27). Sustainable Development of Renewable Energy integrated Power Sector: Trends Environmental Impacts, and recent Challenges, *Elsevier.* https://www.sciencedirect.com/science/article/abs/pii/S0048969722007379.

❑❑❑

SECTION - 7
Hospitality and Tourism Sector

19

Digitalization in the Hospitality and Tourism Sector

Sanath Krishnan Menon
Research Scholar, Manav Rachna University

Dr. Priyanka Chadha
Assistant Professor, Management and Commerce, Manav Rachna University

Abstract

The aim of the current paper is to examine the factors in terms of digitalization which would act as a catalyst in the development of the hospitality and tourism industry. We would also look at these factors in detail and how they are interrelated and are a major stakeholder in revenue generation for the sector. This study is crucial in these times where the customer is reluctant to obtain a service from this particular sector keeping in mind the safety and health issues. The market share of an organization is directly proportional to the kind of modern technologies adopted by the organization and its social media presence. This paper intends to identify the technologies which represent digitalization and also identify the shortcomings. Digitalization is present in every sector and it was high time that the tourism industry too entered the era of digital tourism. As the customer is prone to digitalization from all other sectors, it is very much important to fill that particular void in the hospitality sector. Technologically superior nations have already adopted modern measures and therefore it is very much important for other entities to adopt similar measures in order to stay connected digitally. Over the years certain developments have been made for sure it does require certain amendments and developments. With the progression of Covid-19, customer behaviour and preferences have changed, and cordial behaviour could be mended with proper use of augmented reality.

Keywords: *Hospitality, Tourism, Digitization, Virtual Reality, Augmented Reality, Virtual travel, Chatbot, Artificial Intelligence*

Introduction

Since the inception of ICT (information and communication technology), digital switchover has been a topic of discussion. For the past 20 years, in the hotel and tourism industry, it has been the forte of the conventional agencies to look after the demands of the tourists. It was customary during the older times to visit the agency to get relevant details regarding the destination and tourism related services. It remained a hectic task for an individual with no guaranteed success as the services offered were slow and individuals had to wait for eternity to avail a basic service.

As the time progressed, and with the advent of technology, the demands of the customer changed along with his behaviour which in turn changed the tourism and hospitality industry. As they were obtaining services from other industries, their concept of service also changed with respect to hospitality. People expected services where the organisation was more customer oriented.

As technology has progressed itself, everyone in our family owns one or the other form of digital device. A digital device which could give us a plethora of information, which in turn has affected their decision making and their habits. Unlike the older times you don't need to go to tourist agencies to book a reservation for yourself and instead with a few clicks you could book all the items in your itinerary at one go which would be supported by reviews to back up for. A few years back you would have to call the front office or the respective department to avail service from the hotel, which was indeed tiresome. Now with a few taps on your mobile device or a tablet, you could go through all the services offered by the hotel and avail one for yourself.

We must be aware that digitization of the tourism and hospitality industry is still at its initial phase. There are 2 spectrums of digitization which must be kept in mind, the digitalization which happens in the industry and the kind of digitalization which is visible to the customer. Hotels could save a lot and bring in more efficiency if they invested more on a digital experience. Hotels also haven't provided a user interface to the customer which is compatible with the kind of technology or interface he is experiencing via other industries. There has been a conflict of opinions regarding the reduction of the human workforce as the hospitality industry is a people centric industry and its more based upon the experience it receives from the staff.

There are various advantages of bringing in digitalization to the industry. It brings in more efficiency and also moves along with the lines of sustainable development. It is also responsible for your growth as it brings in repeat customers and you have more market share in the industry as people could look up to you directly instead of relying on word of mouth. If you do have positive reviews, it would act as a domino effect to bring in more business as other customers would too be looking at that review as a point of reference. Digitalization is more about creating an experience for the guest and is not limited to the customer-oriented services. In this new era, it also includes voice automation, light detection and artificial intelligence. Therefore, it is very much important to focus on the infrastructure and design of the organization. Even though we must focus on creating an experience for the guest we should also make sure that we aren't completely dependent on technology as this industry is people oriented. Even though technology should be the bedrock of an organization, it shouldn't sell itself on those lines.

With the kind of breakthrough technology has made, it has made a lot of organizations obsolete as they weren't able to cope up with the demands of the customer. The habits of the customer aren't stagnant and they look out for more effectiveness. We have reached a phase, where the development of the industry is dependent on digitalization.

As Covid-19 had caused a considerable amount of damage to human mankind, the hospitality industry had lost its focus on digitalization. This should be a call for introspection and should be identified as a blessing in disguise. Customers are more inclined towards digital service and are trying to avoid personnel services. customers are looking for possible solutions in the form of digital concierge, in room technologies and much more. There seems to be a thin line where all the components of the industry, collectively are looking for solutions in order to provide a breakthrough. Even though digitalization is in the initial stages of development, there are some sectors which could give an immense boost to the hospitality sector. Some of them are as follows:

1. Internet of Things (IoT)

In the simplest terms, the Internet of things can be defined as a phenomenon where all the individuals are connected with each other digitally with the help of a physical device. These devices are installed with sensors, processing ability, software, and other technologies and are able to transfer data from one device to another.

These devices are able to receive and send data and are able to communicate with each other. This exchange of data helps in creation of a data bank which could be used for research and development. It mainly revolves around the fact that you could control everything within the purview of technology in the hospitality industry using a device i.e. mobile phone. Even though there may be many aspects to this technology, the hospitality industry has limited its use to personalization. The guests are now able to control the ventilation, heating, air conditioning, control the tv or even set the temperature of the water/ sauna in the bathroom. The technology has advanced to the extent that the device is able to identify you using facial recognition and retinal eye scan. If different profiles are created on the same device, the device is able to command a unit to give out personalized service according to user preferences. With new advancements in GPS and Bluetooth, it has led to a new pathway for location-based services and it has led the industry to deliver out a message to a particular customer at the particular time.

This has been better utilized by the restaurants where they send out messages which read promotional offers or discounts, whenever a guest is near the restaurant. Tourist destinations have also utilized this in a way that whenever a customer reaches a particular city, he is provided with all the information regarding the city along with tourist attractions and information about local transport

2. Robots

Since Industrialization, robotization has picked up at an immense pace. But it should be noted that robots in manufacturing industries have only been used for production purposes. In the industry they are required to pre programmed tasks at set intervals of time, but the hospitality industry is a little bit trickier. The main crux of the hospitality industry is being customer centric and to give personalized service according to his needs. Human elements and human emotions play a major role in an industry such as hospitality. The glaring difference is that machine communication is predictable whereas human communication isn't. As discussed above, as we have a large data bank regarding data as to what are the possible different responses for a particular situation, innovation and technology has made a huge stride in creation of robots for the hospitality industry. After the outbreak of Covid-19, people are more sceptical about human interaction and more inclined towards services which offer more or less no contact.

This in turn has led to adoption of technology which is beneficial both for the customers and the employee and in turn providing a customer experience keeping in mind the safety precautions. The main advantage that a robot provides Is that the cab be disinfected after each checkout and in turn providing a safe handover of the room. The robot would be able to do a UV scan of the room which cannot be done by the naked human eye. They are not prone to exhaustion and they in turn provide services with more efficiency. They could deliver service with precision and accuracy which cannot be done by the human kind. However, they do come with certain drawbacks, such as in the long run they may displace humans at work. Another downside to this is the cost of procurement, as they are very expensive in the present market but if the industry considers this as a potential investment where they don't have to pay the employees, it might be very fruitful venture in the coming years.

3. Recognition Technology

The main task of the facial recognition system is to identify the image against the database of images which could authenticate a user through ID verification services. Recognition technology would involve technologies such as facial recognition, retinal eye scan and biometrics. This kind of technology has garnered the attention of the customers and they are in demand. As discussed above, facial recognition would be able to detect a particular user and would be able to deliver tailor service to that particular customer. Hotels would be able to scan the customer at the check in point and would be able to identify if the customer is a repeat guest or not. If yes, previous credentials could be retrieved from the database and would be provided with a hassle-free check in. Facial recognition technologies are also capable of determining the mood of the individual, their gender, their age and so on. This might be able to create a database for the future where this data bank could be used to identify their response regarding services which in turn might be used to develop a trend. Facial recognition might also be used to identify and verify a guest and provide access to his room, gym, restaurants, spa etc. This might in turn help the industry to ditch the key card procedure which might save them economically. Linking this database to the city police station might also alert the hotel authorities regarding the presence of a fringe element in their premises. A new concept of selfie pay has been introduced by payment companies, where the customer is able to pay just after scanning his face. This leads to a hassle-free checkout with minimal human interaction. This could be used at various outlets such as gyms, restaurants, bars and various hotel outlets.

4. Virtual Reality (VR)

Virtual reality is a feigned environment which is created with the help of artificial intelligence and is presented to the user in a way that it deems real to the user. Its main objective is to create a simulated environment using technology and simulate all 5 human senses. It is sure to bring in a competitive advantage to the hotel and give the customer an experience of what he is going to commit to before booking.

VR could be used as a marketing tool where the hotel uses this technology to give a customer a preview of things that are about to come. They would be able to go through the catalogue of service which they could avail and image representation could give assurance about the vibe of the premise. This could also be used for the purpose of brand creation where customers would be able to experience a live demo of things that he could avail in the future. The brand would be able to create a USP for itself where you could project your organization and show differences where your property is better than others. If the guest is able to differentiate between the different kinds of services that he could avail at different price ranges and if a visual representation is being provided, it would eventually lead to upselling of the room which in turn is beneficial for the property.

5. Augmented Reality (AR)

Augmented reality is a phenomenon which works on the line of virtual reality. The difference lies in the projection of the environment where the real world changes in terms of virtual reality but in augmented reality the real-world environment changes in real -time.

With the help of augmented reality, the hotel could provide a real time experience to a guest which would help them to upsell a room and create more revenue. Augmented reality consists mainly of graphic information i.e. if a guest points his smartphone at a map, the map would open itself and offer information such as local places of interests or nearby restaurants.

As PokemonGo was the frontrunner of augmented reality, various hotels too have worked on that line and have developed various interactive games which would garner attention of the children and adults likewise. This also leads to development of their own application which would make the user interface enjoyable. Augmented environments would also project celebrities at the hotel locations which in turn makes the users stay more enjoyable.

6. Chatbots & Artificial Intelligence (AI)

Chatbots offer pre-programmed responses to the question put forward by the customer and this is one of the most used technologies in the industry today. These responses are devised from the database of questions that the guests might or have asked. A common list of questions is being generated and possible answers to those questions are devised and delivered to guests upon request. This means that a particular human representative is not required and that he could respond to much bigger problems

Artificial intelligence should be used more in the back office of the hotel which deals with revenue generation or to devise a management strategy using the data retrieved from customers. This data should be used to devise a promotional strategy and calculate profit and revenue for the organization during the current period and in the long run.

7. Reputation Management

It has become necessary and evident during these times to read a review of the product or a service before going for that product. As such, the tourism industry is a perfect example where people would love to read a review before going to a hotel or a destination. There are certain ways in which you could deal with a customer review.

It should be made compulsory for the guest to provide feedback at the time of checkout where in case of a grievous complaint the management can take necessary decisions regarding the same. It should also be the same when the guest leaves a positive comment where they could be provided with offers and discounts on their next visit. The management should make sure that the review platform is up to date and that the organization should strive to provide customer service

Future Implications

The benefits of this digitalization would be seen over the years as the demands of the customer would evolve along with technological advancements because it is eventually going to provide advantage to both the industry and the customer. Many forebearers of the industry have understood the usefulness in the industry such as positive customer experience, increased efficiency and brand awareness.

As we proceed more towards customer centric support, the industry would be able to provide much better assistance according to the guest profile. Providing a high-quality customer experience would lead to a customer being a repeat guest which could be cashed in by the industry to generate more revenue.

The amount of brainstorming and research that can be done with all the data available in the form of guest preference and their habits could be used further to create an environment which might suit their likes and interests.

It is important to notice that along with technology different branches of marketing are also expanding. One of the major components of marketing is promotions where it helps you to gain consciousness and helps to raise a conversation with the potential guest. It is a psychological procedure which helps you to gain the customers attention, a medium through which you can engrossment about your products and services and to help gain a potential repeat customer. It's also an opportunity to say that the product has been developed with "want satisfying capabilities". It is also noticed that it increases the acceptance speed of products and services. In order to avail maximum customers, it is very much important to tell your potential customers about the latest products and services which are best done digitally. In this time of technology, the organization could send multiple newsletters, offers, discounts and promotions to an infinite number of potential customers with a single click. With this kind of technology, an organization could penetrate into a market with millions of plausible customers which would offer a higher ratio of converting them to customer or repeat customer.

Customers are looking for long lasting experiences which includes a customer centric experience. Gone are the days when customers only looked for food and accommodation. A guest these days would want facilities that require minimum human effort and are compatible with his digital devices. It is time that hotels dive deep into technological advancements and introduce latest technology in order to garner more guests and more than that it requires the tag of a permanent guest from a customer in order to showcase loyalty to a particular hotel. It is imminent for the hotels to introduce technologies so that they don't lose their existing customers to other properties and dissolve themselves.

Conclusion

Tourism and hospitality industry, digitalization is one of the tools that would open up the new array of possibilities for the industry. It would apply modern techniques to an old school approach. Customers wouldn't be able to lose the essence of food and lodging but would be able to experience it more with more added features that they would remember and would be able to blend in their everyday nature. The main aspect of digitalization still remains to connect an individual with all the facilities available in all the industries using a single digital device. Even though there is a large portion of people who estimate that the industry has attained the breakeven point, the fact remains that we are far away from it. Technology and digitalization work in a way where they evolve. Things which were deemed impossible 20 years ago are done with ease now. The main objective is to develop a more sustainable form of technology which is profitable both for the industry and the customer.

Customers' acceptance of new technology has compelled the industry to adopt digitalization. As mentioned above as every industry is inter related to each other, digitalization would help personalize a particular profile which could be later used in other services. The question the customer has for the industry is that I as an individual is able to avail a particular service in another industry using my digital device. So, what makes it difficult for the hospitality and tourism industry to implement it? The kind of experience that the customer experiences and accepts in a positive way compels him to obtain that service in all possible sectors. Earlier when the guest had to go to a particular destination he had to take reviews from people around him or a tourist agency regarding every aspect of that travel such as booking the tickets for the transport, booking the hotel, booking for tourist destinations and had to accept the only choice which was provided to him . Now with the advent of technology it is very much possible to have a mental image of the destination or the transport through which you would be going. You could book the food and lodging according to your preferences. You are provided options in terms of websites through which you could book these facilities and as to what are the pros and cons to each particular facility which would be put forward in terms of reviews by customers who would have first-hand experience of these facilities.

Many hotels have a conglomerate of services wherein they provide an array of services. Earlier we had to go to a travel agent to book our tickets but now it's possible with the help of the internet. Earlier we had to book a taxi to reach a destination, but now we could call the hotel personnel and they would send us a car which is suitable for us and is decided by us at the spot of our choosing. Earlier we had to check in at the front office, now we could use a kiosk to let ourselves in. There are various examples where we were dependent on various elements of the tourism industry but as mentioned above these services are made, are streamlined and are provided by one single unit.

Market penetration would act as a key factor to sustain yourself as an organization in this sector. Existence of a particular clientele is limited to a limited area. In order to cement yourself as a stakeholder of the industry you would need to have a global presence in the market. Data driven innovation would be key to the future of innovation in this sector. Understanding the cultural shift and implementing a technology in a particular area should be a task which should be performed by the administration. Understanding the pulse of the customer is the data which the organization should thrive for because that would eventually lead to development of product or service which would attract that particular customer to your organization.

It might be better for the hospitality industry to embrace these technologies as soon as possible before it becomes obsolete. It is better that the industry embraces these modern technologies in order to stay connected in this digital world. The hotel management students who in turn become future hoteliers need to update the curricula wherein they dispose of the conventional and orthodox operating procedures and get more inclined towards development of modern technologies as the industry demands specialization in this sector. It is these future hoteliers who have a vision for the future with respect to the industry and its time that technology is introduced to them now so that they can develop it later. With the development of the sharing economy, there is a greater emphasis on social well-being as user experience begins to outweigh ownership. This new approach is also reflected in new forms of payment, which can significantly reduce the profits of intermediaries. The new trends do not seem to be problematic in the tourism industry, mainly because the exact costs and revenues in this sector are not yet clear. It is pretty much evident that industry wants to work on the line of sustainable development along with profitability which in turn would create a path for the future of the hospitality and tourism industry.

References

Zeqiri, A., Dahmani, M., & Youssef, A. B. (2020). Digitalization of the tourism industry: What are the impacts of the new wave of technologies. Balkan Economic Review, 2, 63-82.

Kolobkova, V. A., Romanov, A. A., & Frolova, E. A. (2021). Digital technologies in the hotel industry: new prospects for sustainable development. In Socio-economic Systems: Paradigms for the Future (pp. 387-394). Springer, Cham.

Cigniti (2020, Sept 20). The Impact of Digital Transformation on the Hospitality Industry. https://www.cigniti.com/blog/digital-transformation-hospitality-industry/ retrieved on 5th Sept, 2022

Revfine (2020, Dec 20). 8 latest digital trends used in Hospitality Industry. https://www.revfine.com/digital-trends-hospitality-industry/. Retrieved on 5th Sept, 2022

Aguiar-Castillo, L., Guerra, V., Rufo, J., Rabadan, J., & Perez-Jimenez, R. (2021). Survey on optical wireless communications-based services applied to the tourism industry: Potentials and challenges. Sensors, 21(18), 6282.

Tovaglieri, F. (2022). Post COVID-19: What's next for digital transformation?. https://hospitalityinsights.ehl.edu/what-next-digital-transformation. Retrieved on 11th Sept, 2022

Nyagadza, B., Chuchu, T., & Chigora, F. Technology Application in Tourism Events: Case of Africa. In Digital Transformation and Innovation in Tourism Events (pp. 107-116). Routledge.

❑❑❑

Made in the USA
Monee, IL
12 November 2026

994b27d6-5851-4652-99c0-4bf8f4022a26R01